Cities by the Sea

The Nature | History | Society series is devoted to the publication of high-quality scholarship in environmental history and allied fields. Its broad compass is signalled by its title: nature because it takes the natural world seriously; history because it aims to foster work that has temporal depth; and society because its essential concern is with the interface between nature and society, broadly conceived. The series is avowedly interdisciplinary and is open to the work of anthropologists, ecologists, historians, geographers, literary scholars, political scientists, sociologists, and others whose interests resonate with its mandate. It offers a timely outlet for lively, innovative, and well-written work on the interaction of people and nature through time in North America.

General Editor: Graeme Wynn, University of British Columbia

A list of titles in the series appears at the end of the book.

NATURE | HISTORY | SOCIETY

Cities by the Sea

Urban Coastlines in Atlantic Canada

CLAIRE ELIZABETH CAMPBELL

FOREWORD BY GRAEME WYNN

UBCPress · Vancouver

Originally printed in Canada.
Printed and bound by CPI Group (UK) Ltd, Croydon, CR0 4YY
UBC Press is an eBOUND Digital Certified publisher. The epub version of this book meets stringent accessibility standards, ensuring it is available to people with diverse needs.

Library and Archives Canada Cataloguing in Publication

Title: Cities by the sea : urban coastlines in Atlantic Canada / Claire Elizabeth Campbell; foreword by Graeme Wynn.
Names: Campbell, Claire Elizabeth, author
Series: Nature, history, society.
Description: Series statement: Nature, history, society | Includes bibliographical references and index.
Identifiers: Canadiana (print) 20260176192 | Canadiana (ebook) 20260176281 | ISBN 9780774873376 (softcover) | ISBN 9780774873390 (EPUB) | ISBN 9780774873383 (PDF)
Subjects: LCSH: Shoreline development—Environmental aspects—Atlantic Provinces. | LCSH: Waterfronts—Atlantic Provinces—History. | LCSH: Coasts—Atlantic Provinces. | LCSH: Cities and towns—Atlantic Provinces. | LCSH: Atlantic Provinces —Environmental conditions.
Classification: LCC HD319.A7 C36 2026 | DDC 307.1/41609715—dc23

Canada Council for the Arts Conseil des arts du Canada

Canada

British Columbia Arts Council

British Columbia

UBC Press gratefully acknowledges the financial support for our publishing program of the Government of Canada, the Canada Council for the Arts, and the British Columbia Arts Council.

This book has been published with the help of a grant from the Canadian Federation for the Humanities and Social Sciences, through the Scholarly Book Awards, using funds provided by the Social Sciences and Humanities Research Council of Canada.

Parts of this book have appeared as "Hardened Water: The Remaking of a Coastal City," *Coastal Studies and Society* 2, 1 (2023), 112–36, and "Whatever Happened to Pleasant Street? Rediscovering an Urban Shoreline in Halifax, Nova Scotia," *Environmental History* 25, 1 (2020) 134–49.

UBC Press is situated on the traditional, ancestral, and unceded territory of the xʷməθkʷəy̓əm (Musqueam) people. This land has always been a place of learning for the xʷməθkʷəy̓əm, who have passed on their culture, history, and traditions for millennia, from one generation to the next.

UBC Press
The University of British Columbia
www.ubcpress.ca

For Campbell Jesse

You make the world more beautiful.

In the evening Alice sat on her grandfather's knee and listened to his stories of faraway places. When he had finished, Alice would say, "When I grow up, I too will go to faraway places, and when I grow old, I too will live beside the sea."

"That is all very well, little Alice," said her grandfather, "but there is a third thing you must do."

"What is that?" asked Alice.

"You must do something to make the world more beautiful," said her grandfather.

– Barbara Cooney Porter, *Miss Rumphius*

Contents

Illustrations

FOREWORD

On the Waterfront

Graeme Wynn

Four modest cities in four Canadian provinces lapped by Atlantic waters form the heart of this study. This might seem to be a slight foundation for a book, but in Claire Campbell's hands, these storied cities by the sea are well worth wider attention. All have essentially, eighteenth-century foundations. The largest, Halifax in Nova Scotia, began as a strategic garrison town in 1749, and now counts half a million people in its metropolitan area. St. John's, half the size of Halifax, owns a Royal Charter issued by Queen Elizabeth I in the sixteenth century, but for fifty years it was no more than a village serving the seasonal fishery. The trappings of permanence – a garrison, fortifications, churches, a resident population – emerged slowly; barely 10,000 people lived there when Newfoundland's first year-round governor arrived in 1817. The town plat of Saint John, New Brunswick, was laid out in the 1780s to accommodate loyal refugees from the newly independent Thirteen Colonies, and Charlottetown was designated a county town, though one with very few residents, during the survey of Prince Edward Island in the 1760s. Today, the Saint John metropolitan area includes slightly more than 130,000 people, and the greater Charlottetown area houses about 80,000, of whom fewer than half live in the city itself.

Together, Canada's Atlantic provinces count only 5 percent of the national population, and those who live in and study this part of the country have often lamented the marginalization of regional concerns in discussions of the wider realm. But *Cities by the Sea* – concerned, primarily, with the ways in which the shorelines of Campbell's chosen places have changed through time – appears at, and addresses, a moment in which the perils of such blinkered perspectives are increasingly evident. As global heating continues and sea levels rise (driven by meltwater from shrinking glaciers, ice sheets, and polar ice caps and the volumetric expansion of warming ocean waters), cities, towns, other structures, and land uses on foreshore littorals are at risk – worldwide. Along Canada's 226,000 kilometres of coastline, estimates suggest that approximately 600,000 people occupy land at risk from rising waters, and that number may well rise to 850,000 by the end of this century. Because land in some parts of the country is rising (or rebounding from the weight of icesheets) and in others (such as the Atlantic region) it is subsiding, the magnitudes and effects of rising sea levels will vary. In the Atlantic provinces, increases of 146 to 175 centimetres are anticipated. In the greater Vancouver and Victoria areas of British Columbia, estimates stand at 125 to 130 centimetres; north and west of Hudson Bay, sea levels may fall.[1]

Combining deep familiarity with the region and evidence from historical maps to illuminate her characteristically luminous, almost conversational prose, Campbell carries her readers along waterfront streets in the major cities of each of the Atlantic provinces, searching, as she puts it, "for the older shorelines and wetlands lying beneath the burdensome weight of city blocks and coal yards and freeways." Another way of thinking about this is to see it as a metaphorical tracing of chalk lines on pavements (as Campbell's students at Dalhousie University used to do materially) to show the levels to which water will climb (unless there are significant efforts at mitigation) as sea levels rise. This is an arresting gambit, intended to animate engagement with issues of much wider significance than the simple question of how much land (and infrastructure built upon it) will be submerged. Among these: how past, present, and future weigh on the minds and choices of individuals, and their societies; the effects of "inertia" (the "burden of the past") in shaping land-use patterns and prospects for their modification; the challenge of balancing what people have with what they want and need; and the different means by which we seek to represent, understand, and effect change in our material environments.

Campbell's thoughtful work also lays challenge to the long and widely held conviction that urban environments are *sui generis,* that they are somehow set apart from Nature because they are the products of human action and design. Eighteenth-century English poet William Cowper encapsulated this belief, embraced by nineteenth-century Romantics and echoing into the present, in a single line – "God made the country, and man made the town" – in a discursive piece of blank verse exalting evangelical faith and ennobling humankind.[2] In the pages that follow, Campbell brings us to recognize that Cowper's felicitous formulation embodies two meanings of fabrication: "creation" and "untruth"; that neither of its constituent phrases stands without qualification; and that jeopardy lurks in continuing fealty to notions that, unbendingly, set humankind apart from nature.

Few of these general claims will be entirely new to those familiar with the recent literature in environmental history. The putative origins of this field of study are often traced back to the publication, in 1864, of *Man and Nature,* by George Perkins Marsh, subtitled *Or, Physical Geography as Modified by Human Action,* in which the author proclaimed his intent "to indicate the character and, approximately, the extent of the changes produced by human action in the physical conditions of the globe we inhabit; to point out the dangers of imprudence and the necessity of caution in all operations which, on a large scale, interfere with the spontaneous arrangements of the organic or the inorganic world."[3] Elaborations on, extensions to, and refinements of this mandate have been legion in the century-and-a-half since Marsh wrote.[4] Indeed, there is now a sizeable shelf of books, not all by historians, that focus directly on the massive topographical, hydrological, climatological, and floral and faunal changes brought about in the course of city building.[5] Alongside them are several others, noted by Campbell, that deal specifically with changes in the watery littoral zone of cities developed on lake and ocean shores. Many of these offer impressive accounts of hills flattened, swamps drained, rivers canalized (or culverted), and new land made (by extending foreshores) to accommodate the trappings of industry, commerce, and transport.

Today, Singapore, Hong Kong, Shanghai, Doha, and Dubai are spectacular exemplars of a process that has continued for centuries, and been massively accelerated by the ready availability of enormous amounts of investment capital and the inexorable power of brute force technologies. But they are only the most celebrated, and promoted, of the projects that have, by one recent account in the journal of the Royal Institution of Chartered Surveyors, added 2,500 square kilometres to coastlines around the world since the turn of the millennium.[6] Some two-thirds of what were once tidal flats skirting the Yellow Sea have been reclaimed. China, the Netherlands, and South Korea lead the world in making foreshore land, but not all of this is turned to urban uses.[7] The United Arab Emirates, Bahrain, and Qatar together include almost as much land taken back from the sea as the much larger United States of America. Beyond the Netherlands, most reclamation projects were on a relatively small scale until the middle of the twentieth century. For all their experience and ambition, even Dutch engineers were unable to reclaim most of Lake Haarlem (183 square kilometres) in the Netherlands until the refinement of steam pumps in the mid-nineteenth century; the Zuiderzee drainage was not completed until 1968.[8]

Successful reclamation projects have enabled economic expansion, sustained urban metabolisms, and become integral to city fabrics through the decades. Ocean-lapped colonial towns that grew into cities invariably filled in or extended their foreshores. Downtown and portside areas of such major centres as Cape Town, South Africa; Wellington, New Zealand; and Rio de Janeiro, Brazil, all stand on fill. Toronto, on the shore of Lake Ontario, Canada, serves to illustrate a process repeated, time and again in broad terms, in most other port cities. There, the Toronto Harbour Commission erected an imposing new building in 1917. Located on the end of a wharf extending into the harbour, it had water at its feet. At its back, railroad tracks and marshalling yards, built on landfill in the nineteenth century, filled the five hundred metres or so of new land that separated the building from the original shoreline of Muddy York, along Front Street. Today, a busy four-lane road runs by the sidewalk in front of the Harbour Commission site; Lakeshore Boulevard and Scotiabank Arena have replaced many of the tracks; and condominium and office towers ensure that nary a glimpse of Lake Ontario is visible from the front steps of 60 Harbour Street, now about half a kilometre back from the water. In each of Cape Town, Wellington, Rio, and Toronto, areas that once hummed with the industrial, commercial, and transportation activities that tied ports to ever more extensive hinterlands have been remade as spaces of recreation, leisure, and domicile. Sports stadia, entertainment venues, museums, restaurants, galleries, and high-cost high-rise residential towers climb above parks and promenades on waterfronts everywhere, as societies that once saw foreshores in overwhelmingly utilitarian terms now embrace the new aesthetic of consumption.

Canadian folk singer Stan Rogers marked this transition in Nova Scotia as early as 1976, when he sang of the "Upper Canadian concrete and glass" that had replaced the workaday world of "Fisherman's Wharf" on the Halifax waterfront.[9] By his account of the waterfront, the old song – "haul away and heave her home" – no longer echoed through the rigging of fishing schooners unloading their catch. Old ways had been traded for new; men who understood winds and currents had given way to "a single tide of tourists passing through." Few but romantics, ready to ignore the cold winter ocean and the gale's "awful song," lamented "what's done and gone," sang Rogers – even as he was led to wonder about the future, and "what it's coming to." *Cities by the Sea* is animated by the same question.

Because sea-level rise is a looming, omnipotent menace that haunts the future – and the pages of this book – as it threatens the continuing utility of the mass of infill and pavement along the urban foreshores at the centre of *Cities by the Sea,* it is instructive to compare Campbell's work with a couple of recent projects intended to raise awareness of this threat on Canada's east and west coasts. Both of these recent endeavours are computer- and team-dependent initiatives, heavily funded and widely publicized. The east coast work emerged from a desire to minimize risks to the economy and ecology of Prince Edward Island by preparing for the impacts of climate change. It is based in the UPEI Climate Lab, which developed a broad *Prince Edward Island Climate Change Adaptation Recommendations Report* for the provincial government, emphasizing practical ways of reducing vulnerability, increasing resilience, and benefitting from climate change opportunities.[10] CLIVE (the CoastaL Impacts Visualization Environment) is a tool providing analytical visualizations of coastal erosion and future sea-level-rise scenarios.[11] Its use of videogame technology allows users to "fly" around a 3D representation of the Island using a controller. They can select different erosion and sea-level-rise scenarios and observe the environmental changes that ensue from any perspective at any scale. Touted as intuitive and immersive, CLIVE has garnered national and international attention for its effectiveness in educating citizens and raising awareness.

Geovisualization is also at the centre of a cluster of similar west coast initiatives, exemplified here by the Future Delta project.[12] Developed at the University of British Columbia, in conjunction with schools in the municipality of Delta at the mouth of the Fraser River, it is an interactive 3D game in which players move virtually through digital representations of their communities and make behavioural choices that generate alternative climate scenarios, the flood risk impacts of which are reflected in detailed simulations of Delta neighbourhoods. Or, in the language of its developers:

> The year is 2100 and you live in a dystopic future where the City of Delta has failed to address climate change adequately – it has been affected by sea-level rise, storm surges, heat waves, & fires, and communities are facing food shortages, high-prices, increased traffic and pollution. Travel back in time to present day and see if you have what it takes to change the future ... for better or for worse.[13]

These 3D interactive simulations ask what the future holds and seek to build understanding of climate change, in hope of catalyzing community support for efforts to mitigate the effects of sea-level rise.[14] Their role in building awareness is often emphasized.[15] By showing water coursing inland, flooding cars and streets, surrounding familiar landmarks, and enveloping the houses of viewers, these representations bring people to see the effects of rising sea levels in local, immediate, and consequential terms. Proponents of visualization insist that seeing climate change matters. In this view, making an abstract concept tangible, and showing how its results impinge on individuals, encourages individual and community efforts to avoid future scenarios envisaged in the model. After a series of meetings introducing CLIVE to eight Island communities in 2024, CBC News reported that the "climate change simulator tool" drew gasps and tears from residents.[16]

In the second quarter of the twenty-first century, the chances of any book eliciting similar responses, or providing the dopamine hit of 3D visualization games, seem scant. But different media convey different messages, and *Cities by the Sea* reinforces this McLuhanesque maxim in important ways. Aware, as historians tend to be, that "the devil is in the details," Campbell makes good and deft use of various interpretive lenses to unfold her accounts of changing urban form in her chosen Atlantic cities. None of these places (or indeed their waterfronts) is discussed exhaustively. Each, rather, illustrates a particular conjuncture of time and circumstances. Together these accounts span about 150 years, between the last third of the eighteenth century to the first quarter of the twentieth, but different questions, concerns, and approaches shape the discussions of each city. In Saint John, the focus is on the foundational transformation of an extensive foreshore (where a major river meets the large tidal surge of the Bay of Fundy) into a commercial harbour. Here, a close reading of the city charter, a document built on precedent to shape an uncertain future, reveals a persistent tension in city building: the reconciliation of private interests with the public good. St. John's, by contrast, illustrates encroachment on the harbour by "made ground" and the ensuing unintended consequences that greatly vexed the harbourmaster in the 1870s and 1880s. Charlottetown, truly a town for much of the nineteenth century (its population increased about twenty-fold in that span, but it barely topped 11,000 in 1911), offers a different story. There, the tidal "sea creeks" and adjoining wetlands that squeezed the original town plat were largely sacrificed (by reclamation and infill) to the self-same imperatives of growth and development that gave shape to much, much larger urban places, in what nineteenth-century clergyman and historian Robert Vaughan termed "the age of great cities."[17] A century on, Charlottetown remained small, a city barely deserving of the modern connotations of that word, but a place on a human scale. Halifax, again by way of contrast, was remade as its capacious harbour was envisaged as the nation's leading eastern port. As investment poured into improving the efficiency of the "functionally magnificent" port of Halifax, the city of Halifax was increasingly detached from the working waterfront. This trajectory was (literally) cemented by construction of the "mammoth" Ocean Terminals early in the twentieth century. And as Campbell points out, the nineteenth-century mantras – consumption, expansion, national interest – that sustained this growth remain current in the twenty-first century.

Cities by the Sea is less a comparative history than a series of vignettes that tease out abiding themes in the processes of urban development in Canada's Atlantic provinces and beyond. These sketches remind us of the burdens of the past – from alterations to the material environment to the shaping influences of precedent and experience – and their capacity to constrain thought, frame the ways in which choices are made, and limit options for the future. They also drive home the material manifestations of words and ideas, as they mark environmental transformations brought about by rhetorics of progress, expansion, and advancement. Much of Campbell's story is divined from the rich archival store of historical maps of the northeastern foreland of North America, many of which illuminate these pages and all of which Campbell uses to good and uncommon effect, by asking what they show of landscape change, and seeking to identify the persistence of former landscape features in the four cities today. Insights garnered from all of this are further polished by Campbell's innovative framing of each chapter around different sources: the (dry) legal charter incorporating Saint John; the St. John's harbourmaster, whose grumblings and inclination to seek remediation rather than regulation revealed unforeseen consequences and aligned with the preferences of those interested in improvement, productivity, and profit over limits and containment; PEI newspapers, for debates that spilled beyond the water of the Governor's Pond and the East Marsh to broach issues of race and class and revealed the lingering inertia of sunk capital in Charlottetown; and literary sources that provide insight into human aspirations and sentiments as citizens struggled to find meaning in the remaking of the landscape, and the transformation of space into place (or vice versa) in Halifax.

Clearly, this book does more than simply chronicle the alteration of waterfronts. The chapters that follow illustrate some of the many different ways in which people have sought to represent, understand, and effect change in their material environments, and the powerful shaping effects of ideology and rhetoric upon public discourse. What emerges from this account is a view of earlier urban waterfronts not as "past scenarios" but as palimpsests – as multi-chaptered stories of artifice and intervention, shaped over centuries, in which the forces of persistence, novelty, conversion, abandonment, and renewal vie against one another in response to the siren songs of progress and improvement, and often regardless of environmental limits and the societal and other consequences of change. Ultimately, Campbell asks her readers to think deeply about the appropriate balance between what people have, what they want, and what they need.

This is a crucially important invitation, at a time when rising anxiety about global heating and rising sea levels is characteristically met with "business-as-usual" responses. Governments at all levels express concern at the implications of environmental threats but typically incline to short-term or temporary fixes (aka "preparedness plans" and physical adaptations). Risk mapping, to identify the magnitude of the problem and design efficient responses, is often the first step in responding to climate change concerns. And like most first steps, it shapes the path taken. Worried about heat domes in certain areas? Target subsidies for the acquisition of electric heat pumps to provide respite from soaring temperatures (and, possibly, reduce CO_2 emissions) in places vulnerable to extreme heat. Vexed by the destruction of private property by forest fires? Develop a Wildland Urban Interface (WUI) Risk Class Framework and use this to guide investment in wildfire risk reduction. Apprehensive about the effects of global warming? Electrify the transport system to reduce fossil-fuel consumption. Nervous about the threat of coastal inundation? Retrofit, raise, and re-inforce existing dykes – or build new ones.[18] Although discussions of the "Future Delta" project allude to the importance of moving to "more sustainable lifestyles," reducing carbon emissions, and "exploring new ways of living – being, knowing, and doing," the optional future scenarios offered to those who play this game are, revealingly, building a seawall, engaging in managed retreat, creating offshore barrier islands, or moving to floating homes.[19] These too are adaptations, efforts to alleviate the effects of sea-level rise rather than address its cause. They treat symptoms rather than tackle the malaise.

Responding to her own invitation to consider the balance between desire and need, Campbell turns, in the final chapter of this beguiling book, to urge the importance of doing things differently – of bucking the stream of tendency that has made urban waterfronts, the cities they serve, much of the material world, and many forms of life on earth increasingly vulnerable to environmental risks (such as heat, fire, wind, and flood) heightened by climate change.[20] In these final pages, she calls upon the power of the educated human imagination to contemplate the past, not simply as what happened but as a repository of disparate ideas and mixed outcomes; to find discordant notes amid its dominant refrains; and to dream of (or imagine) what might be.[21] Focusing on Georgetown, Saint Andrews, and Lunenburg, three towns whose origin stories differ little from those of the three larger counterparts in the Maritime provinces examined in previous chapters, Campbell wonders why their growth faltered and whether there are lessons to be learned in and from their histories.

Today, these three places are marked by compactness, and they mostly eschew the growth ethos embraced by larger places. At one level, Campbell sees these towns as what some in this troubled, crumbling world are beginning to call "Islands of Sanity," or "parts of the system that hold resilience when the rest of the system comes out of balance."[22] Such enclaves, the story goes, are relatively safe places, with a robust sense of community in which human lives are lived in closer connection with the rest of nature, and creativity and kindness flourish. They suggest the possibilities of lives lived apart from the strife and woe, the pressures, expectations, and detrimental environmental consequences of modern techno-industrial society, and its seemingly insatiable desire for more of everything. Still, questions remain. Can Lunenberg, Georgetown, Saint Andrews, and other places like them, hold on to the attributes that seem to make them sanctuaries from a "world too much with us" and remain viable?[23] How might the advantages that such places seem to offer be extended to far more people than those who currently live in them? Is it possible to remake larger places in their image? Do such attributes as a human scale and a strong sense of community erode as populations grow? Does everyone regard these characteristics as benefits? These are difficult questions yet to be resolved. Campbell's foray into these towns, and what she takes them to offer, is preliminary, tentative, exploratory; evocative rather than definitive. At best, it is partial. As Campbell notes, "finding the foreshore" might "help us imagine healthier, more resilient cities," but there is harder work to do if increasingly urgent and necessary changes are to happen.

The prosperity and lifeways of Canadians, like those of most people in the so-called Global North, are heavily dependent on inanimate sources of energy. On a per capita basis, Canada has one of the highest rates of energy consumption in the world; annual averages are almost triple those of the European Union, and approximately half of Canadian consumption is generated from fossil fuels. Modern-day Canadians find it difficult to forgo the comforts and benefits provided by their command of energy. In a vast land of cold winters and great distances, these have become integral to warmth and connection. Although much has been said about the importance of the energy transition – and a "fair" transition at that – Canadian efforts to reduce emissions from the use of fossil energy sources have faltered. In several other countries, the situation is even more grave, as politicians bow to expedience, energy corporations resist constraints on production and profits, and most people define well-being in materialistic terms. Filled with hubris, reluctant to recognize ecological limits, disinclined to forgo the Western world's centuries-old commitment to growth based on the extraction of Earth's endowment, and averse to curbing appetites for technological innovation and consumption, the globe's most prosperous nations have long overtaxed the Earth's biospheric endowment and overloaded its ability to absorb the waste of ever-expanding populations and economies. Since 1970, this pattern of ecological overshoot has increased steadily; by 2026, humanity's demands exceed Earth's renewable biological capacity by about 75 percent; by 2050, the human footprint will probably exceed what Earth can sustain by a factor of two. Humankind – and the world – face the consequences, in the form of global heating, rising sea levels, and other disruptions of Earth's biogeophysical systems.

Averting the impending catastrophe implied by these broad metrics will be immensely difficult. Canadians live, today, in a highly visual culture. The old adage about a picture being worth a thousand words may never have seemed more apposite. But this exchange rate, and the fast-paced market-society given to short-term perspectives in which it thrives, is far from fixed. The price may be right in certain circumstances. In others, it is surely open to question. Although the number of words in *Cities by the Sea* is far greater than the number of illustrations, their relative value is not to be determined by simple arithmetic. Pictures are integral to this book, but they are incorporated into thoughtful, nuanced stories to develop an important argument, elaborated with imagination and verve. When Campbell writes – with a nice turn of phrase and characteristically lucid insight – that "'land artificially won from the waters' is not as secure a possession" as those who extended foreshores, filled swamps, and paved greensward (if not paradise) in the nineteenth and twentieth centuries chose to believe – she does more than make a point about human hubris. By drawing attention to difference, context, and process, Campbell shows what a book can do.[24] She provides a platform of understanding from which to reflect on the past and imagine a different future. More than this, her book illustrates how the study of history (and by extension, the humanities) can contribute to the vital task of addressing urgent current crises, by prompting readers to reflect on blithe, often imperfect, assumptions, such as the conviction that prosperity and progress can be extended indefinitely and in defiance of ecological limits by human ingenuity and technological intervention. *Cities by the Sea* nudges us to recognize the unsustainable ecological costs of the ways we live now and to begin the truly hard work of recalibrating our definitions and expectations of "the good life." This is important and absolutely necessary labour.[25]

Acknowledgments

Writing this book has been wonderful, and difficult. Since 2013, I have lived in central Pennsylvania, over a thousand kilometres from the border at St. Stephen, New Brunswick. Thinking and writing about Atlantic Canada has kept me there in some ways even as I live here, like Jane Stuart of Lantern Hill, "because in a very real sense Jane was still living on the Island."[1] (And I'm not even from the Maritimes; I have to admit I grew up in Toronto. Sorry.) I'm here and away, landlocked – and a bit at sea.

But it was also a way to address some of the frustration and anxiety of living in a human-authored climate crisis. I wrote parts of this book on days it was too hot for my son to play soccer outside. Parts of Nova Scotia and New Brunswick burned while I wrote. For many of us, environmental history isn't an academic exercise; it's born of love of and concern for a place. A constant *but what can I do* thrums in the background. Does writing about asphalt and artificial coastlines help? Likely not, but it's *what I can do.*

Selfish as those motivations are, I am full of gratitude for those who made this possible. Chief among them are archivists and librarians across Atlantic Canada, including Simon Lloyd and Donald Moses at UPEI; John Boylan at the Public Archives and Records Office of PEI; Peter Larocque, Michelle Hébert, and Jennifer Longon at the New Brunswick Museum; Joshua Green and Mary-Ellen Badeau at the Provincial Archives of New Brunswick; David Mercer, Linda White, Colleen Quigley, and the Centre for Newfoundland Studies at Memorial University; Maureen

Peters at the Rooms; Paula French at the City of St. John's Archives; Alison Froese-Stoddard with the Halifax Public Library; Elena Cremonese and Susan McClure at the Halifax Regional Municipality Archives; and John Macleod, Anjali Vohra, Rosemary Barbour, and the reference desk at the Nova Scotia Archives. Thank you also to Enrico Moretto, who photographed insurance maps in Toronto, and many others who answered innumerable emails and questions. When I was in graduate school, I had a terrible picture of "the archives" as silent, windowless rooms (an image coloured by the '70s brutalist architecture of college campuses and government buildings) ... but these are delightful places filled with people as keen on your project as you are, who *literally* know where the bodies are buried, and who are endlessly patient and helpful across the miles or in person. Thank you, all.

No one plans to get rich being an historian, so you're grateful if you don't have to go into (too much) debt doing it. This project was made possible through the generous support of the Association for Canadian Studies in the United States, the Clements Library at the University of Michigan, the Ocean Frontier Institute, and at Bucknell, the Provost's Subvention Fund, the Scadden Fund, and the Humanities Research Centre.

The *process* of writing can be solitary – large chunks of this were written at 4:30 a.m. – so it helps when you can connect to a community. A special thanks to Harry Holman and an extra-special one to Edward MacDonald, the guide of all things Island. Thanks also to the CJAM group and NEAR-EH, who read parts of this; and to the anonymous reviewers who read the whole thing, as did my editors at UBC Press: James MacNevin, Katrina Petrik, and especially Graeme Wynn, as much for innumerable conversations over the years as for his foreword here.

Lastly, my thanks to Anthony and Campbell Jesse. I may still look for things that aren't there anymore, but I see everything differently because of you. Including soccer.

Cities by the Sea

INTRODUCTION

The Vanished Shoreline

The sea is come up upon Babylon: she is covered with the multitude of the waves thereof. Her cities are a desolation, a dry land, and a wilderness, a land wherein no man dwelleth, neither doth any son of man pass thereby.

– Book of Jeremiah, 51:42–43, KJV

We have built cities by the sea for millennia. But with climate change reshaping terrestrial space and redrawing the lines between land and water, these cities face uncertain futures. As the world's population congregates ever more at the water's edge, ever more in cities, amid ever more tumultuous seas, the situation – and the history – of our coastal cities becomes ever more important.

With sea levels rising, with flooding and storm surge growing more intense and more frequent, it can be a genuine hazard to live without knowing where the shore is – or more precisely, where it should be. Several years ago, historian Andrew Lipman wrote about living in New York City in the fall of 2012, when Hurricane Sandy tore up the eastern seaboard of the United States. Power was cut, the skyline went dark, and rising water revealed the shape of an older city. "Flooding on Manhattan," Lipman remembered, "exposed the island's long-lost face, as the high waterline traced the contours of the seventeenth-century coastline." I have been haunted by this image. It was, as Lipman noted, a temporal collision, between the city's historical geography, its contemporary functions, and the promise of future disasters to come.[1]

Nor was it unique. Catastrophes of the new climate era expose the fissures in our urban fabric and every year demonstrate how ill-prepared cities laid out in the eighteenth century and built out in the nineteenth are for the demands of the twenty-first. One of the most engineered cities in the world, New Orleans, founded in 1718, will forever be synonymous

David Wilson, *By the Sea at the End of the World* (St. John's, Newfoundland, 2022). | Courtesy of the artist

with the disaster of Hurricane Katrina in 2005. In 2024, Toronto, laid out in 1793, flooded to a standstill as a rain-swollen watershed reemerged in subway tunnels and across highways. And this is not limited to the New Yorks and Torontos of the world; climate-related disaster is a feature of modern urban life on any scale. People in Atlantic Canada talk about storms of record and memory – Hurricane Juan (2003), Hurricane Fiona (2022) – that paralyzed cities in the region for weeks. As assemblages of mass and permanence, of brick and concrete and asphalt, cities may seem purely objects of human design. But climate change is fracturing that façade, bringing water to the surface, exposing the vulnerabilities of what Jared Orsi termed "the hazardous metropolis."[2]

Cities by the sea have special reason to worry, because sea-level rise calls into question their very existence. As environmental historian Ted Steinberg writes, "The sea is knocking at the doors of coastal cities all across the world."[3] But we've rarely acknowledged that we have *made* our cities more vulnerable to those stormy seas by building out, into, and over water. Indeed, we have been allowed to forget the extent to which our cities were built within and because of the place of water and other natural features. Today's waterfronts bear little resemblance to those of even fifty years ago.

When we lived in Halifax, a mid-sized city on a small near-island of a peninsula, university students used chalk to trace the lines of future sea levels along the sidewalks of Upper Water Street – which, of course, had been the city's original waterfront.[4] I began to wonder how much of the harbour's edge that we can see – rail yards, shipping berths, parking lots, ferry terminals, boardwalk – was really shoreline, or where the shoreline really was.

How did we come to this?

Progress and Precarity

Cities by the Sea asks why our cities look the way they do, and how these choices have positioned us so precariously in the present moment. The cities considered here occupy different kinds of coastal environments, ranging from a sediment-rich estuary on the Bay of Fundy to a steep cut of shale and sandstone in the North Atlantic. But they, along with many others, share a similar biographical arc, as they mirrored, embodied, and enacted larger transformations that have remade the earth over the past two centuries. Courses of fresh water and sheltered harbours that invited colonial settlements were absorbed and transformed in the construction of industrial ports. The curves of natural shorelines, in tidewater or salt marsh, were aligned and filled to delineate water and land and thus impose civic order on littoral space; to repackage shoreline as property; to defy spatial constraints; and to make new ground that could support the infrastructure of fossil fuels. It is this ground that is now the front line of coastal flooding.[5] Carried by the prevailing ideological currents, and propelled by new energy regimes, these transformations enabled people to excavate, convert, or compromise a plurality of ecosystems for human need and preference. Over time these interventions profoundly marred our sense of where water is or might be, and they left us unprepared for the environmental changes we now encounter in ways that the narratives of progress and nation-building have never acknowledged.

Cities by the Sea searches for the older shorelines and wetlands lying beneath the burdensome weight of city blocks and coal yards and freeways.[6] In seeking the water's edge, we excavate through a long genealogy of choices and decisions made in pursuit of security and prosperity; what those in the eighteenth and nineteenth centuries called "improvement" and those in the twenty-first might simply call "development." As Graeme Wynn has observed, "progress – not restraint – was the watchword of the Victorian era; few believed then that there were limits to growth." Nature's riches were to be acquired and applied for material gain in ways that, as Wynn further reminds us, have benefitted many of us.[7] Today, though, we are compelled to acknowledge the costs of these decisions, what historian Donald Worster has called the "chains of environmental consequence," either unforeseen by or unconcerning to our predecessors.[8] As Lauren MacDowell put it, our task is now to confront "the grave consequences of the development ethos as it played out throughout Canadian history, not to condemn, but so we can begin to develop strategies to create a livable, sustainable environment in the future."[9] An environmental history of coastal cities thus asks us to consider thoughtfully and critically such language of development where we find it today and the assumptions that still govern our attitudes toward the natural world, on the one hand, and societal well-being, on the other.

But as we will see in the chapters that follow, some of our predecessors were aware that coastal expansion and "made ground" were not without their complications. In 1864, American naturalist George Perkins Marsh identified "land artificially won from the waters" as evidence for his argument that "man has done much to mould the form of the earth's surface." While primarily concerned with agricultural reclamation and drainage projects that encompassed a larger area of land, Marsh considered harbour and coast improvements that "invaded the realm of the ocean" on a significant scale of "mass and weight of matter" as belonging to the same genre of human impact "in their physical and their moral bearings."[10] Later, British geologist R.L. Sherlock detailed other examples of "adding to the area of the country," whether in silting up estuaries, draining marshland, or otherwise filling in the foreshore (iron slag, he reported, supplied a particularly solid foundation for building). Sherlock gave special attention to the artificial expanse of the Liverpool docks, a port particularly important for Atlantic Canadians.[11] The ambiguous, mutable character of the foreshore – "neither true land nor sea," or what Sherlock called "debateable territory" – would be overlaid with a stratum of modernity as shapes of streams and ponds and tidal flats became ghostly silhouettes in concrete and asphalt. In Atlantic Canada, people expressed concerns about harbour health ranging from silt and pollution to safety of navigation as more and more of the waters were enclosed.

Such concerns have found a new platform as discussions of sustainability, climate change, and the Anthropocene have entered public parlance.[12] At once dynamic, permeable, and essential to human survival, coastlines are emblematic of the ecological precarity we face in the new climate era. We know that sea levels are rising, albeit unevenly, because of melting ice and the expansion of warming water molecules. The *rate* of sea-level rise is also increasing, also unevenly; thanks in part to processes of land subsidence, Atlantic Canada and the Beaufort Sea will see the fastest rates of sea-level rise in the country. The sites considered in this book are predicted to experience sea levels rising from seventy-six centimetres to well over a metre by 2100. Water Streets in Charlottetown, Saint John, and Halifax will indeed be under water. Harbour Drive in St. John's will be *in* the harbour.

Scientists are now also predicting a permanent alteration of the world's thermohaline circulation systems, or in simpler terms, the ocean currents. Storms that gather energy from warming air and water are increasing in speed, intensity, and frequency. As seawater crashes on shore, it heightens flooding, coastal erosion, the loss of barrier islands, and saltwater intrusions into freshwater aquifers. Rivers, too, rise to bursting, swollen from storms and caught in urban constraints; Saint John, for example, experienced record levels of "freshet" flooding in 2018 and 2019.[13]

These are the ripped-from-the-headlines stories that we associate with coastlines and climate change, but they too have a past. Modern parking lots and petroleum tanks are the clearest expression of the "Great Acceleration" of anthropogenic change from the mid-part of the twentieth century, but as we will see, there is a much longer history leading to this point. The challenge for scholars is to balance our intellectual and ethical responsibilities: to identify as precisely as possible the historical actors, to craft language that best ascribes degrees of historical responsibility ... while remaining focused on *why* such work matters. The inescapable reality is that human actions are degrading the planet. This is not an academic question.

Water, Cities, and the Contribution of Environmental History

Part of the problem is that we have given ourselves permission to forget.[14] This allows us – in a New-Year's-resolution kind of way – to focus on forward-looking promises of innovation rather than the costs of past decisions. The term "sustainability," after all, suggests maintaining a place or a practice into the future; "adaptation" implies making the best of present circumstances. We are already accustomed to measuring environmental change in the language of scientific data and statistical accountings: areas of sea ice, global temperatures, or wildlife populations. But a historical lens is essential to understanding how we arrived at these numbers, to addressing and redressing situations that originate in the ancient, imperial, and industrial pasts.[15] Environmental history considers those origins: how nature informed human choices and actions and how, in turn, these actions affected the natural world that surrounded them. It might document physical or material interventions, like pouring concrete wharves or building railroad bridges. Or it might explore expressions of imagination and ideology, asking, for instance, what signifies progress in the industrial era. In a way, environmental history asks how people have understood and responded to the possibilities and constraints supplied by the environment around them.

Despite our mental picture of cities as compositions of cement and asphalt, cities play a special role in understanding this historical relationship with the natural world. As Stephen Mosley observed, "As urban centres have become the main battle-grounds where sustainability and climate change are concerned, the need for a historically informed understanding of the interconnections between cities and their environments – and especially between cities and the sea – has become increasingly urgent."[16] Cities are the places where, statistically, most people live, and they are the largest drivers of carbon emissions. I would add that cities are also places with significant clusters of historic properties that explain how we got to this point, and where we can tell that story in immediate and effective ways. Notably, city waterfronts were crucial sites for sustaining the expansion of capitalism. It was along the waterfronts that expensive and lasting infrastructures, designed to capture distant markets and circulate wealth, could be constructed. Nineteenth-century cities spoke in the language of destiny; exerting control over water, in particular, would help determine their place in history. Yet at the same time, these cities – and especially their waterfronts – experienced cycles of construction and devalorization in ways that I suggest map onto changing energy regimes.[17] Today, those cities embody that inherited and ongoing tension between a historical desire for ordered linearity – water*front,* coast*line* – and the space and movement of water itself: a tension made manifest in every photo of storm waves striking a wharf or boardwalk.

Knowing how we acknowledged water in the past – as border or intrusion, asset or inconvenience – and how people responded to each characterization is essential to understanding a city today. Not surprisingly, scholars have generally focused on the larger metropoles and megacities, like London, New York, or Los Angeles: places where the demands on water have been most pronounced, the infrastructure has been most expansive, and the greatest numbers of people are experiencing the effects of climate change.[18] Although, as Matthew Booker has noted of San Francisco, even here people are often oblivious to the city's coastal infrastructure and the natural features on which it sits.[19] Most such cities share a story of waterfronts moving from sail to steam, from industry to recreation, and sometimes, to projects of restoration. There is a homogenizing quality to modernity: the processes of industrial transformation have made the ecologies of cities more alike, overwriting distinct natural features with similar networks of fossil fuels and water supply. Ironically, "revitalization" or gentrification can have a parallel effect, with the generic boardwalk for tourists.[20] But cities encompass a spectrum of responses to water and other environmental forces, ranging from sublimation to accommodation and adaptation. Frequently, industrial cities took shape by walling off saltwater or burying fresh water, but at other points they conceded the persistence and presence of water. There is something to be said for balancing the grand narratives of global force – capitalism, industrialization, and so on – with the local, tangible dynamics of street, wharf, or neighbourhood.[21]

There has been a wonderful surge of scholarship exploring the relationships between cities and water, showing how the organization of water undergirds the entire organization of urban space, visible and invisible, domestic and public. Much of this has focused on fresh water, in cities both ancient and modern, especially the management of what Joel Tarr called the urban metabolism: the movement of water through human use, in supply and disposal, and related issues such as sanitation, public health, access, and municipal governance. The topography of watercourses within cities also explains much of a city's spatial organization and social inequities.[22] Scholars of port cities, in contrast, tend to emphasize the traffic on salt water, the imperial networks of commerce and government, and the displacement and projection of industrial demands onto coastal and nearshore areas.[23] Tendrils of water, fresh and salt, bound these cities to larger hinterlands. If we look to where fresh meets salt, we find ourselves in the emerging field of coastal studies, which recognizes the value of knowing how we have acted *by* if not *on* the sea, in ways strongly aligned with environmental history, by exploring aqueous spaces including estuaries, islands, and wetlands. In this book we will often encounter attempts to corral both fresh and salt simultaneously: filling streams, ponds, and saltmarshes while fixing the contours of tidewater. As Wilko Hardenberg writes, such "terraqueous history" connects individual locations to a global coastline, a dynamic all the more pressing in the Anthropocene.[24] Indeed, communities in Atlantic Canada have long been deeply invested in their port connections within the region, the "Boston States," and Great Britain. It is worth considering the ways historically networked cities have shared ecological characteristics and linkages.[25] With the most amount of coastline of any country in the world, Canada has much to contribute to the international conversation about cities, coastlines, and environmental change.

But in Canadian history, the east coast tends to fade from the national narrative following the turn to continental expansion, from sea *to* sea, leaving those Atlantic cities to an older era of tall ships and wooden wharves. Indeed, it may be hard to picture the eastern seaboard, with its iconic fishing sheds and lighthouses, as agents of industrialization and engines of climate change. But just as Richard Judd argued for the importance of the American northeast in environmental history, because of its centuries of record spanning cycles of depletion and recovery, Atlantic Canada is likewise an incredible archive of environmental change.[26] The region has witnessed some of the greatest environmental disasters in Canadian history, including one of its largest oil spills with the wreck of the *SS Arrow* in 1977; its largest hazardous waste site at the "tar ponds" in Sydney, Nova Scotia; and the collapse of the largest commercial harvest in the groundfishery closed in 1992. It remains a flashpoint of the frictions of colonialism and reconciliation with ongoing tensions between settler and Indigenous fishers over access to stocks. Now it is the site of some of the most aggressive evidence of climate change, as the beaches of Cavendish and foundations of Fortress Louisbourg alike erode into the sea. And its cities embody fundamental themes and patterns in urban history, from colonial plats to condominiums, from public parks to buried rivers.

But the writing of environmental history in Atlantic Canada (and arguably elsewhere in Canada, too) has skirted its cities, leaning toward rural landscapes of resource use or tourism.[27] By way of illustration, Thomas Raddall's *Halifax: Warden of the North,* first published in 1948, remains the most popular and best-known history of a Maritime city. The region has a strong tradition in historical geography, which made the study of regional landscapes an important question but which tended to study rural communities and resource lands.[28] In the later twentieth century, many historians were concerned with explaining what they saw as the inequities resulting from Confederation and the preoccupation with continental expansion. In their interpretation, the Maritimes were excised from nation-building projects (except as a coal supplier), their natural resources increasingly

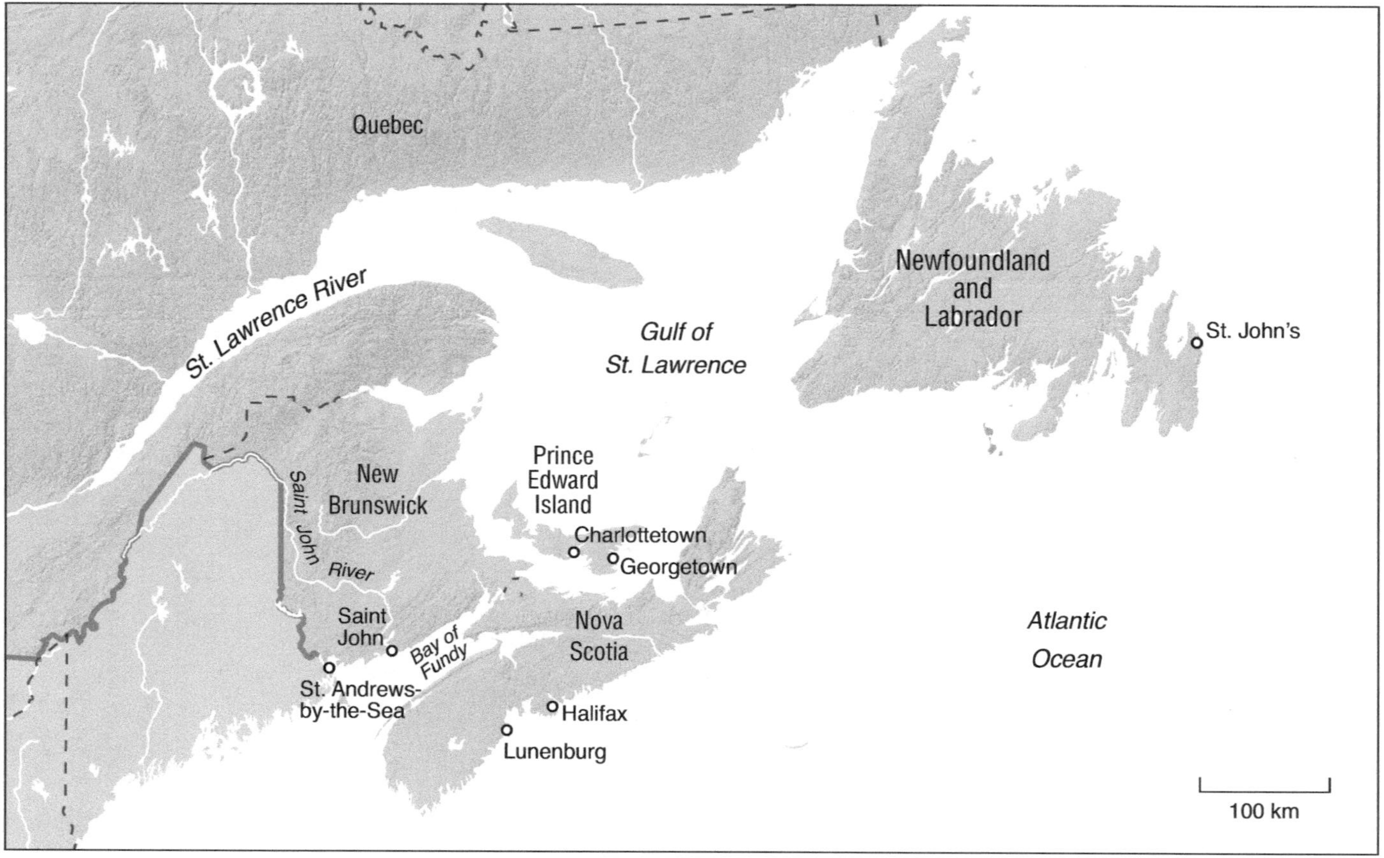

Map of Atlantic Canada. | Cartographer Eric Leinberger

controlled by central Canada; wilting from outmigration, and reduced in political influence in a federation tilting west and north, the Atlantic provinces were now at a disadvantage because of their location by the sea. The "Acadiensis School" saw Atlantic cities as hinterland capitals at best, artifacts of a golden age confined to the era of wooden ships.[29] While environmental history in the region has expanded to consider such topics as tourism and environmentalism, it has inherited the geographical interests in landscapes and the leaning toward political economy.

If, however, we situate Atlantic Canada in a narrative of environmental change, we see different things again. For one, the language of growth and extraction, or Canada's "development ethos," was not confined to prairie grain and northern bitumen. Behind the picturesque saltbox lies a concurrent commitment to seeing the Atlantic horizon as a frontier and to romanticizing colonization and extraction. As historian Arthur Lower wrote exaltedly in 1953, "And as to the sea! ... Is it not a frontier, too, which calls out all the resources and ingenuity and adaptiveness of man?"[30] As important, pushing past the conventional curricular "break" of Confederation in 1867 finds threads of purpose running from the late eighteenth century that we would recognize in the twenty-first. It allows us to draw on the exciting work from scholars of early America and reminds us to consider spaces of power other than those inscribed by imperial and later nation-state borders.[31] And most relevant to *Cities by the Sea*, it finds coastal development as a manifestation of certain ideas of progress, growth, and "improvement" pervasive in the region and beyond for centuries.

The Invitation of Cartography

In February 1746, *The Gentleman's Magazine* of London quoted the preeminent French cartographer Jacques-Nicholas Bellin as saying, "Geography is so necessary to illustrate history that they ought to be inseparably connected."[32] *Cities by the Sea* takes Bellin at his word. In rediscovering older shorelines, this book relies heavily on maps, plans, and surveys. In speaking to public audiences, I have found historical maps are both popular and resonant sources: different enough to be intriguing, visually alluring to the point of art, and imprecise (by our estimation) enough to be entertaining. They are wonderful expressions of environmental history, whether older ideas about nature or past interventions in the landscape. At the same time, the prospect of reading, let alone questioning, a map often comes as a revelation to my university students, few of whom have ever held a paper map or atlas. They know maps only through the ubiquitous Google, where water is a single matte shade of blue and Starbucks are located more precisely than shorelines.[33] For them a map is an unexamined digital convenience, and the presented geography a fait accompli.

But places and their representations alike are historical texts, human constructions, the work of human hands intended for human use. As Brian Harley recognized, maps inscribe the practices and priorities of their creator and their era. "Each map argues its own particular case," as he wrote.[34] Cartographic literacy is, therefore, an important skill: the ability to read maps critically; to understand the relationship between their spatial, visual, artistic, and textual elements; and to situate these records in historical context. Though we think of modern cartography as a scientific practice of measurement and data (not unlike our standard expectations of the other environmental fields), any map is an artifact of its time, a product of historical forces at work: ambition, diplomacy, epistemology, commerce, and technology. They are authored by forces of politics, capital, and belief, as well as (if not sometimes more than) science or curiosity. Maps speak through symbolic means such as the images of the cartouche or the inscription of new place names, as well as material record; they are a medium of description, persuasion, and projection.[35]

In a sense a map is a nonsensical proposition, especially in the conventional forms most familiar to us. The idea of inscribing a changeable, irregular, living ecology into a flat, fixed form, from a position above the earth, then claiming to know or understand a place, owes much to the reason of the Enlightenment and the territorial ambitions of European (and American) empires. As Matthew Edney observes, the result is that we think we can "take hold of the map as an artifact, turn it whichever way [we] desire, and physically dominate it ... The physical domination of the map promotes the intellectual conviction that one dominates the depicted world."[36] But it is the ability of maps to convey both the acquisition or creation of knowledge – pieced together – and the statement of purpose *for* that place that makes them so valuable for seeing how people have understood nature in the past. As Nancy Coulin and Carola Hein explain in their study of port cities, mapping is "an act of design: orienting, navigating, and in a state of becoming ... cartography can capture both activity and desire [and] can narrate stories as well as delineate property and relate scientific facts."[37] It is thus worth asking why any map works the way it does. What are the conventions that render unfamiliar places intelligible – what Adam Grydehøj calls "geographic legibility"[38] – and why are these accepted? What is the "case" that each map is arguing? What is included, and what is left out? How have we made littoral space knowable and valuable to a terrestrial audience?

It is crucial, then, to recognize that maps are created for specific purposes. Maps record not only the process of examining or claiming a new environment but also plans to transform it. I would volunteer a gentle aside to M. Bellin that maps do more than illustrate history; they help enact it. They are unusual as sources because they can relay the passage of historical time and visualize a chronology of past, present, and future.[39] They annotate older interventions or artifacts of who was there and what was built. They document current occupants or existing practices and occupations. But they also declare – and try to legitimate – further claims and agendas. Coastlines in particular offered locales for "an assertion of power and an exercise in the production of knowledge," and both – power and knowledge – enfolded local ecologies into imperial and then industrial expectations.[40] Maps thus align with the intellectual, political, economic, *and* structural architectures arrayed along the waterfront. As we see water lots or bridge rights-of-way drawn, we see the quest for possession and profit, clarity and order. And yet, within these delineations there are hints of older features, like wetlands, in place names and vacant spaces. In short, for all its claims to authority and surety, a map is as complex and complicated as the history it depicts. Maps show history and nature, alike in motion, being made and remade. Especially with urban shorelines, which can be entirely rebuilt in a few decades, they invite us to ask: Who was here? What did they think of it? How did they live here? How did they change it?

In its pursuit of vanished shorelines, *Cities by the Sea* draws on a wide variety of maps and cartographic images, dating from the eighteenth century to the present day. With each we need to consider not only what (or where) the map is showing us but how it does so and the choices of scale, perspective, composition, and purpose. Survey maps and town plans, for example, are often prescriptive or anticipatory, assigning space in ways that privilege certain types of land (or water) use over others. Harbour charts, reassuring in their precision, emphasize depths, tidal reach, and navigational hazards. Insurance maps detail streetscape fabric, prioritizing commercial activity and property lines. But since the goal is to understand the place and time that shaped these maps, each chapter also draws on a different textual or documentary record to complement the cartographic biography. For Saint John, it is the city charter; in St. John's, reports from the harbourmaster; for Charlottetown, the city's newspapers; and from Halifax, literature and fiction.

Writing from Away

This is in large part because I am, to use a Maritime expression, from away. Since 2013, we have lived in a small college town in central Pennsylvania, two days' drive from the border of New Brunswick. Then, in spring of 2020, the COVID-19 pandemic closed in and threw all of our usual academic practices into disarray. Classes pivoted to online platforms, meetings were held by the hour on Zoom, libraries and archives closed, family care crashed into our work lives. Of course, researchers often have to write from away. I teach with a geologist who studies glaciers in Alaska, a botanist who collects plants from Australia, and a literary scholar working on texts from eighteenth-century Suriname. The liberal arts require us to both contemplate the ground on which we stand and look far to the horizon. But the pandemic revealed some of the assumptions of academia and the invisible requirements of research: personal autonomy, professional security, the financial resources for extensive travel. And as the climate crisis intensifies, we must acknowledge that our trips to the archives come at a cost of carbon.

There are some things we can do differently. Thanks to expanding digital collections, we can access more sources than ever before; we can scrutinize rare and fragile documents, linger over details, move between collections and locations with an ease that otherwise would be impossible. I can read and write about places a thousand miles away, using records from Paris to San Francisco. I owe an enormous debt to the online resources supplied by projects like Island Imagined at the University of Prince Edward Island, the Leventhal Map and Education Center at the Boston Public Library, the Digital Archives Initiative at Memorial University, and Canadiana Online. Indeed, I use these digital collections every day, in teaching and research. I don't consider myself a digital humanist or digital historian; I simply want to be able to *read* things from where I am, whether that is an eighteenth-century map of Saint John or a twentieth-century redevelopment study of Halifax. And yet, writing this book from away has made me simultaneously grateful for digital resources and aware of their limitations. Walking around a city after poring over its maps, I am always struck by how disoriented I feel, caught between the representations I have studied and the *feel* of the place. I have been intellectually prepared to know street names and cardinal points but can't know how the wind tunnels at street corners, where the sidewalks slope. The map, like any historical record, still can tell only part of the story.

Cities by the Sea

This book follows the making of four of the most important cities in Atlantic Canada: Saint John, New Brunswick; St. John's, Newfoundland; Charlottetown, Prince Edward Island; and Halifax, Nova Scotia. Each has a slightly different temporal focus, but lapped together they span what we might think of as a long nineteenth century, reaching from about the end of the Seven Years War to the Great Depression. Of course, there are significant differences between these four cities, their harbours, and their colonies-turned-provinces. Two are primarily sedimentary estuaries, two are ocean-facing rock; two are on the mainland (at least for the moment), two are islands. Still, there are a number of themes that thread through these cities and tie them to larger themes and concerns of their contemporary world.

All were agents primarily of Anglo-British empire, sited on harbours used for centuries by others, notably Mi'kmaq, Wolastoqiyik, and French. As planted towns they were agents of imperial claims and nodes of connection in the networks of expanding commodity frontiers, the emergence of which defined Eurocentric capitalism.[41] As entrepôts each was bound in different ways and through different resources to a wider hinterland. Each evolved a form of the "mercantile triangle" sketched by geographer Martyn Bowden, with bands of activity – shipping and wholesale, financial and retail, community institutions – stretching back from the waterfront.[42] (And each would expand that waterfront to maximize desired use.) Shorelines were subsumed and catalogued by the mandates of national and transnational movement and the premise of growth, with the waterfront transformed into a "geographically ordered, complex, composite commodity." As geographer David Harvey influentially argued, sinking (or "imprisoning") capital within the built landscape constructed a "second nature," a familiar concept in environmental history.[43] At each place a remarkable natural harbour was redesigned – amplified, then compromised – by its transformation into a port, competing for the position of "eastern gateway" from sail through steam through cruise ship. Each city's waterfront was deeply affected by the change in energy regime, as the "forest of spars" of sailing ships gave way to the hardened infrastructure of fossil fuels. These waterfronts are now fossils of these older regimes themselves.[44] But without an accompanying historical explanation, we can forget how these came to be.

Each city altered its fresh water as well as its tidal lines, folding watercourses into urban fabric in ways that never quite erased them. Sometimes buried, sometimes drained, creeks and brooks persisted in shaping urban neighbourhoods, a reminder that port cities, while ocean-facing, are sited within watersheds. Each decision reflected contemporary debates about private interests and public good; whether the littoral should be a water lot or a city commons; and what *kinds* of benefits could accrue from a waterfront location. The enclosure of waterfront for commercial and then industrial uses, the expansion of technological scale, and the continuing projects of artificial shoreline have confined public access to narrow waterfront corridors.

Today the anthropogenic waterfront telescopes the challenge of climate adaptation for any kind of viable, postindustrial city. The problem is that we are trying to reorient places that were not designed to be sustainable; to retrofit motive and purpose on a hardened shoreline. While storms and sea-level rise might suggest a "strategic retreat" from a hazardous shore, it seems unlikely any level of government or corporation would concede to abandoning the enormous amount of sunk capital or, by extension, the ethos of growth.[45] Indeed, clusters of condominiums are now wedged in the unused pockets of space. We know that greener spaces are valuable for everything from public health to environmental resilience, but is it possible to remake or reclaim the shoreline from beneath the weight of industrial infrastructure?

At its heart, *Cities by the Sea* is about how we have altered our waterfronts in defiance of limits and the environmental consequences of this history. It begins at the mouth of the Wolastoq or the Saint John River, using a 1761 chart as an environmental baseline, a measure of the harbour before the arrival of thousands of Loyalists and the transformation of tidal flats into a commercial entrepôt. This chapter reads legal and legislative documents, principally the city's charter (itself modelled on ideas imported from New York), to see how municipal authority was framed in relation to harbour space. The city was made the "conservator of the water," imagining a future where "commerce smoothed the rugged strand."[46] By the middle of the nineteenth century, though, the environmental costs and complications of intense development in the port and resource harvest upriver were becoming apparent in the amount of pollution choking the river and tide. The particular contours of the flats and estuarine harbour invited such amounts of infill that Saint John's harbour would experience the most dramatic physical transformation of any city on the British-Canadian seaboard.

In contrast (and only slightly confusing), the older St. John's, Newfoundland, grew slowly compared to other seaboard cities. But it, too, had

George Hubbard, *Pownal Street wharf on the Charlottetown waterfront* (1849). | PEI Museum and Heritage Foundation, HF.74.12.4

ambitions for transformation and growth. Pinned to the steep north slope of the harbour, Water Street formed a thin and chaotic commercial artery that embodied both the town's precarity in the North Atlantic and its dependence on maritime harvests beyond the Narrows. This chapter traces how a coastal city expanded its working waterfront even within geological confines of the harbour to create usable space for shipping traffic. Expanding the foreshore allowed St. John's merchants to bring the wealth of the ocean within the city's ambit. Infill laid the groundwork, literally, for more ambitious, irrevocable transformations amid the turn to fossil fuels, including the addition of a dry dock and railway complex at the head of the harbour. But the extent of made ground sparked debates over public water and private property – debates recorded by the harbourmaster's investigations in the 1870s and 1880s into "harbour encroachment."[47] The waterfront was where the wealth of the sea came ashore; whether the Grand Banks fishery or oil on the Scotian Slope, the architecture and infrastructure of the waterfront has expanded in pursuit.

Next we travel to the island dubbed the Garden of the Gulf, but instead of the pastures and apple orchards of Avonlea we visit the capital, Charlottetown. This is the story of how a place initially praised for its

aqueous features, a sheltered harbour and navigable "sea creeks,"[48] gradually lost the two wetlands in its downtown core. To the west of the town, the idyllically named Spring Park Brook typified the ways nineteenth-century cities placed multiple demands on their fresh watercourses, from the industrial to the aesthetic. In particular, the ability of water to function as an urban partition was symbolized by the situation of the Governor's residence on one bank of the brook mouth and the Black community known as "the Bog" on the other. Ultimately, proximity to park and suburb protected a fraction of the brook as a civic asset where it emptied into the harbour, though much of the rest of it was buried. On the eastern edge of the town, a coastal marsh – unevenly described and understood as "land covered by water"[49] – fell within the sights of the new railroad and its associated complexes of bridge and building. Hardening the marsh with fill and paving supplied a convenient if drab urban periphery but erased memory of the marsh itself, leaving it a perplexing anomaly on the outskirts of the city's downtown.

The last chapter profiles the city of Halifax and how the contours of a world-class harbour were remade "for the improvement of the trade and commerce of the whole Dominion."[50] Seventy-five years after Samuel

Cunard launched a transatlantic steamship service from Liverpool, the federal government began one of the largest infrastructure projects in Canadian history in constructing a railway/shipping nexus on the south end of the peninsula. The construction of the Ocean Terminals (along with an oil refinery across the harbour in Dartmouth) meant masses of earth detonated and rearranged for and by the fuels of industrial modernity – forces made apparent by the Halifax Explosion in 1917. To construct and represent such an "assembly of enormous and potent apparatus,"[51] visual sources begin to excise humanity, whether in engineering schematics or aerial photography. Literature brings us back to earth, to human scale and experience, so to complement the visual records this chapter turns to three of Canada's best-known writers of the twentieth century – Lucy Maud Montgomery, Hugh MacLennan, and Thomas Raddall – all of whom lived in Halifax in the early part of the century and who write about the city in evocative and revealing ways.

And then, this book asks: Does it have to be this way? Is it possible to live by the sea in other ways? How else have we done so? Atlantic Canada has hundreds of smaller coastal communities whose picturesque and endearing façades, with streets of wooden saltboxes in jellybean colours, belie the fact that they, too, shared the nineteenth-century imagination and aspirations for growth. But they did not all become ports and centres of industry with harbours of cement. By way of epilogue, we visit three Maritime towns that were laid out for settlement in the mid-eighteenth century and enjoyed conventional industrial prosperity in the nineteenth. But unlike the four major cities profiled earlier in the book, Lunenburg in Nova Scotia, Georgetown on Prince Edward Island, and Saint Andrews (by the Sea) in New Brunswick stayed small. Though each faces its own challenges *as* a small community, from tourist congestion to economic lethargy, they have different relationships with the coast. They invite us to imagine what it might mean to live with different choices and different kinds of security and prosperity. They may offer some different possibilities, experiences from the past, and other ways forward through this new climate era.

In 1545, Jacques Cartier presented to Francis I an account of his voyages to "les isles de Canada," or the islands of Canada.[52] Five hundred years later, as the seas rise, our coastlines return to a world of islands. How will we respond?

1
Harbour Improved
Saint John, New Brunswick

A city so rapidly increasing in wealth, population and extent, and possessing in an eminent degree, capability and advantages which only require to be more fully developed to place St. John in the first ranks of the Commercial Cities of this Continent.

– *John Harvey, Lieutenant Governor of New Brunswick, 1838*[1]

Prologue: 1761

Engineer Robert Bruce was posted to Annapolis Royal, Nova Scotia, in the summer of 1761. As the Seven Years War ground to an end, the British Admiralty had directed special attention to charting foreign harbours and ports, especially along the extensive shorelines of North America.[2] Bruce was to evaluate the state of forts in the colony of Nova Scotia, which at that point included present-day New Brunswick. And so here he was, on the other side of the Bay of Fundy, trying to make military sense of tidal flats. Accordingly, Bruce centred his plan of the Saint John harbour on Fort Frederick, built three years before on the site of the older French Fort Menagouèche and – although there is no mention of it on this plan – millennia of Wolastoqiyik (Wəlastəkwewiyik/Maliseet) travel. But in the end, the precise, star-shaped fort is almost imperceptible on Bruce's plan, a mere pushpin holding the plan against the river mouth. Bruce conveyed the real force of the landscape in the sweeping contours and vast extent of tidal mud, estuarine marshes, and lines of shoal. It was a portrait of the life of a shoreline where one of the largest and most powerful river systems on the Atlantic seaboard meets the highest tides in the world.

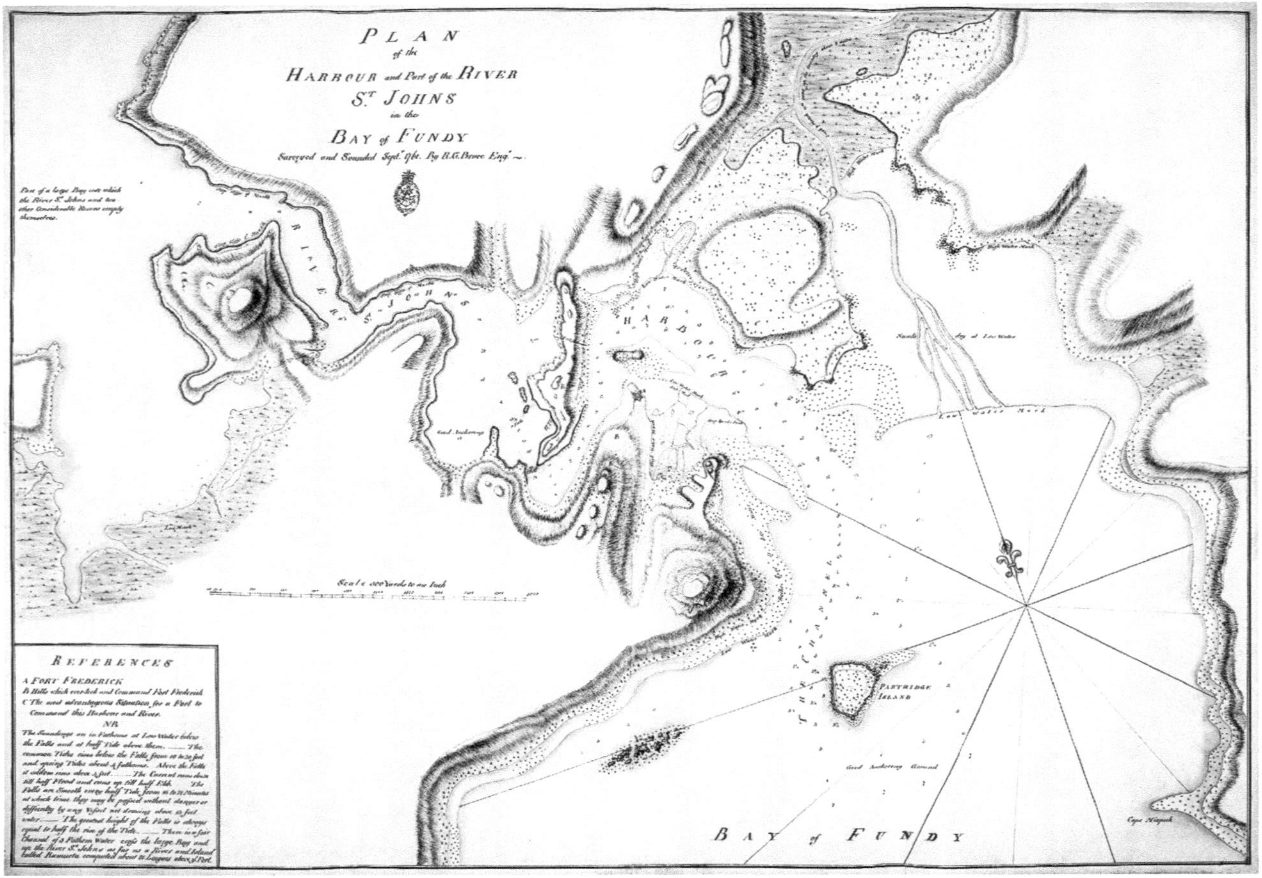

Robert George Bruce, "Plan of the Harbour and Part of the River S.T Johns in the Bay of Fundy" (1761). | Norman B. Leventhal Map and Education Center at the Boston Public Library/British Library Collection

Like other mapmakers of his era, and as the Admiralty instructed, Bruce was primarily concerned with ensuring safe passage for sailing ships. Harbour charts frequently included numeric depths and soundings, as well as visual cues in landmarks, sea-view profiles of the shoreline, and written instructions for navigating safely into port. Bruce's chart had all of these. Soundings outline the channel of greatest depth, with anchorages noted in the inner harbour and above the falls, shoals marked "X" and with cautionary annotations ("Rocks bare at Low Water"). A description of the tides includes directions for how best to ascend the now-famous "reversing" falls, when half-tide brings the ocean level to the river for a few minutes. Though preoccupied with the enormous extent of tidal flat, Bruce did give some attention to the hills surrounding the fort. (As the French had discovered at Fortress Louisbourg, coastal defenses encircled by high ground could prove to be a mixed blessing.)[3]

An Environmental Baseline

Bruce's chart tells us a great deal about what the shoreline at the mouth of the Saint John River may have looked like before decades of infill. Most importantly, it reveals how little of what today we think of as the city of Saint John was there. Most space in this map is turned over to the shapes of shoreline: *Low-water mark – soft mud – Steep Gravelly Bank – Sands Dry at Low Water.* Portages by foot across narrow necks of land were still possible. This was an amazingly convoluted and dynamic tidal space, much of it occupied for moments or hours by water, fresh and salt; neither land nor sea. What is most astonishing about the view from the twenty-first century is how much of this is gone, buried beneath constructed waterfront.

By 1850, Saint John would emerge as one of the largest cities in British North America and one of the most important industrial ports on the eastern seaboard.[4] It would also engineer the most remarkable transformation of an urban harbour on Canada's east coast, part of what historian T.W. Acheson described as "the spread of the material city."[5] Over the course of the nineteenth century tidal flats and estuary pools were shaped into anchorage and timber ponds and then sunk beneath railway wharves and grain elevators. Most of what Robert Bruce marked as tidal wetland is now enormous amounts of pavement; indeed, the shapes of asphalt match almost exactly the contours of tidal flat and marshland on Bruce's plan. This is new, artificial ground, built almost exclusively for industrial infrastructure powered by fossil fuels. It is this ground that will be underwater again amid rising sea levels.[6]

This is, then, in part, the story of how a harbour became a port. A harbour is a natural feature, albeit with particular value for human uses. But a port is a human project: a geography of infrastructure and movement determined largely by economic ambition.[7] Ironically, the making of the port can overwrite the features of the harbour that encouraged such ambitions in the first place. What historians can do is unearth the shape and allure of the original harbour before the artifice and extension of the port, and identify the moments and motivations for "increasing in wealth, population and extent" that drove such a transformation even by the end of the eighteenth century. It is worthwhile to examine Saint John in its earlier decades to understand the course set by what one contemporary called the city's "large conceptions and lofty anticipations."[8]

"Large conceptions and lofty anticipations" could characterize most nineteenth-century settlements and state-building projects, of course, but Saint John deserves special attention because of its harbour. This is, frankly, a complicated place. The Wolastoq, renamed the Saint John River, is one of the most powerful and extensive rivers in eastern Canada; it empties into the Bay of Fundy, home to the highest tides in the world. This is a place of exceptional water power, flowing down from the forested uplands of the interior and up from the wider Atlantic. The range of tide exposes an extraordinary amount of ocean floor. Unlike smaller freshwater courses in cities like Charlottetown and Halifax, the Saint John River is central to the existence and experience of the harbour and could not be buried, redirected, or manipulated in the same way. Saint John also had a complicated political landscape: the segmented geography of the harbour was overlaid with civic and commercial interests, local and away, all jostling for influence.

An Unkind Geography: Historians and Saint John

Given the standing of Saint John in nineteenth-century British North America, there is surprisingly little scholarship on the shape of the city. Historians have instead tended to use the landscape to explain political character. Often this meant a backdrop to a narrative of pioneer heroism, a faceless but menacing antagonist of wilderness, or the means of enormous extractive wealth for industrial entrepreneurs. "Geography had been notably unkind to Saint John," wrote David Bell in 1983, echoing the initial reaction of the Loyalists two centuries before: "There was little in the way of man-made improvement to encourage the notion that the land was habitable."[9] That the harbour is judged "unkind" is, of course, more a reflection of Loyalist expectation, and the resulting narrative of deprivation and suffering, than natural capacity. The forests of the upper Saint John River valley were no less intimidating in their extent but at least promised security and wealth, in land grants and then the timber industry that would dominate New Brunswick's economic identity.[10] The city's geography has also been used as a setting for political theatre, as a way to demarcate civic divisions between authorities of the Upper Cove and labour at the Lower Cove, between the governing Common Council and the Chamber of Commerce, and later, between corporate actors such as the railways and the federal government.[11] But there is little discussion of the stunning physical transformation of the harbour *by* those actors. Indeed, there has been more written on railways than on the harbour, even in the making of the city that claimed to be Canada's winter port by the end of the nineteenth century.

So it is worth exploring the making of that port earlier in that century: the ascendancy of the city and the precedents of the industrial era. For this we have a unique historical source in the city charter issued in 1785, followed by legislation compiled in the British North American Legislative Database, to help us understand how Saint John and colonial New Brunswick understood and imposed upon this complex and dynamic geography. The charter framed the legal responsibilities of the municipality in the harbour, but also pointed to the city's sources of revenue; the legislative record reflects the major concerns of an aspiring commercial metropolis.

Eighteenth-Century Saint John

Saint John, like Halifax and Charlottetown, is known in large part for its origin story. All three cities were planted and peopled to shore up the British presence on the eastern seaboard during the turbulent eighteenth century – either before the Seven Years War (Halifax) or during it (Charlottetown), or after the American Revolution (Saint John). And in all three cases, the story is not nearly that simple. Certainly, the Loyalist narrative suggests that Saint John sprung fully formed in response to the arrival of the first ships from New York in the spring of 1783: "Instant cities created through a Government fiat."[12] Thus was supplied the grist for two centuries of popular history:

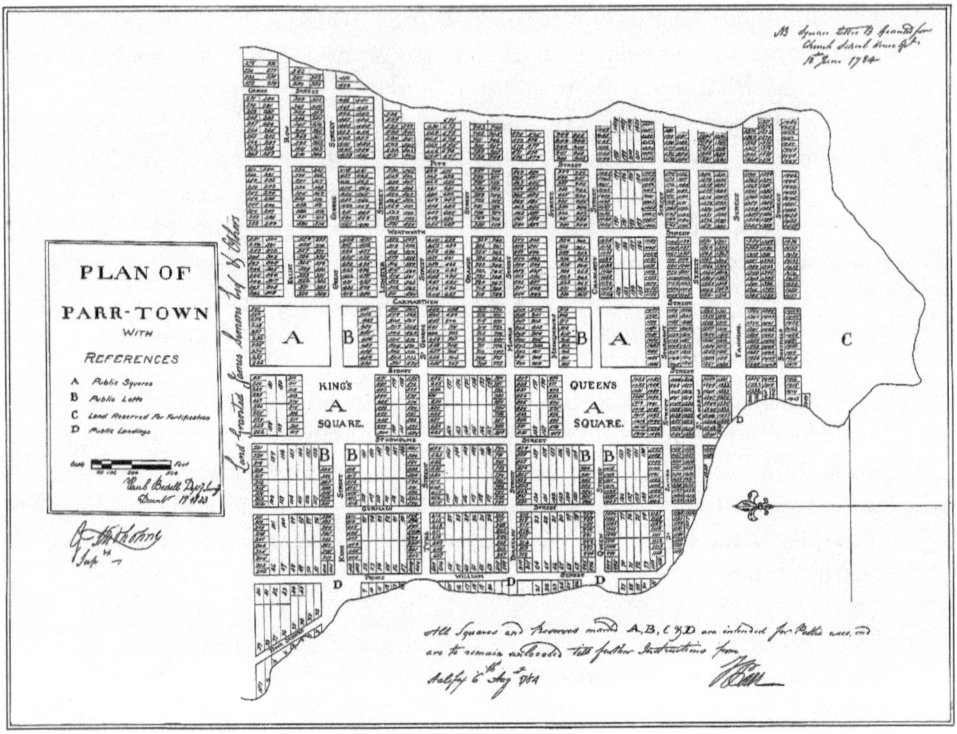

Paul Bedell, "Survey of Parr Town" (1783). | Provincial Archives of New Brunswick RS656-10-35-4

> The indomitable perseverance and strong spirit of men, who for the first time braved the horrors of a bitter winter, in miserable hovels, in a dreary region, struggling with difficulties and privations incalculably severe, surrounded by dismal swamps and endless forests, rather than yield one iota of their duty to their sovereign and their country![13]

The geometric town plat laid out by surveyor Paul Bedell that December promised to mirror and reinforce a self-image of imported and dignified social order, its tidy alignment suggesting a resolute stand against the chaotic hostility of the Patriot cause on one side and northern wilderness on the other. Like other "model towns" planted along the seaboard after the Seven Years War, Saint John's singular and repetitive grid was the product of military art and imperial design. There was no acknowledgement of coastal topography; no concession to the indelible slate bedrock, the steep hills tumbling to the harbour, or the variable tides.[14] Instead, the

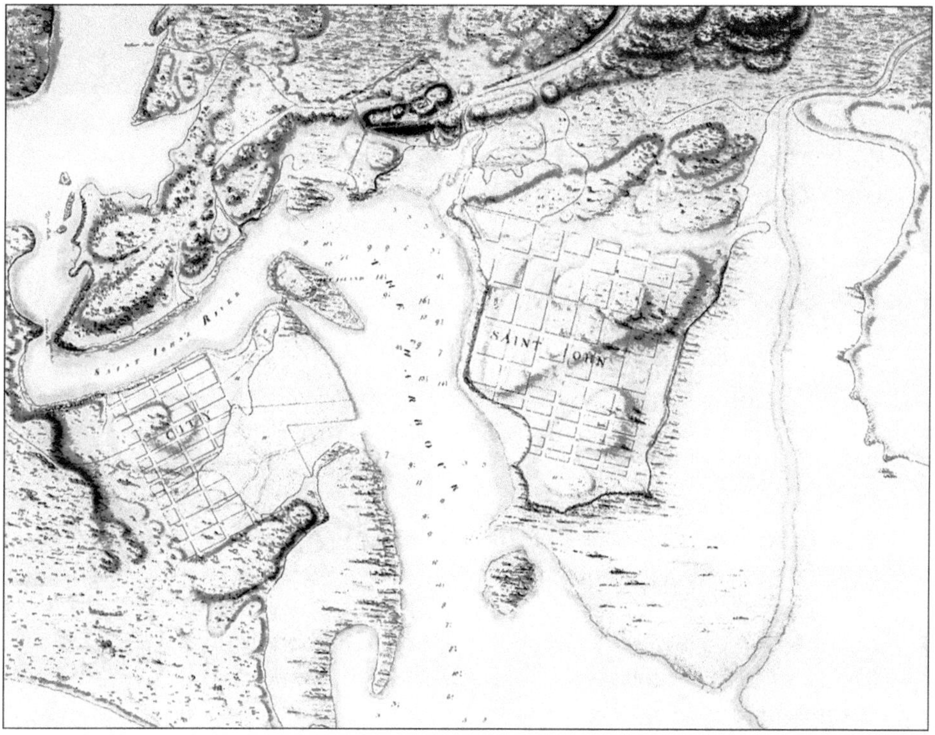

William Lambton, "A plan of the Town and Harbour of Saint John, in the Province of New Brunswick from an actual survey taken in the year 1785." Note the large areas ("H") on both sides of the harbour, including enormous amounts of tidal flat, "Reserved to the Crown at present in the possession of the Mayor and Corporation." | Library and Archives Canada, H1/240/St. John/1785

plan supplied hundreds of lots to be assigned as property to refugee claimants, preserved several points of public access to the waterfront, and inscribed imperial names along the streets to soothe the Loyalist soul.

But as we know from Bruce's plan, the mouth of the Saint John River was neither unknown, unoccupied, nor unwelcoming. Most extensively, the Wolastoqiyik had cultivated and harvested the river valley for centuries. The portages Bruce recorded on this plan were presumably learned from local Wolastoqiyik. The river bounded the territories and movements of other members of the Wabanaki Confederacy, including the Peskotomuhkati (Passamaquoddy) and Mi'kmaq. The river mouth had attracted French traders as early as the 1630s and became a site of contentious rivalries in both commerce and military strength between French and British through the seventeenth and eighteenth centuries. Just as Planter and Loyalist settlers benefited from occupying intervale lands cultivated by the Wolastoqiyik, the city of Saint John was built on top of, making use of, dykes from the French era (also visible on later maps).[15] Indeed, a small trading operation on the Portland shore was provisioning those Anglo-American settlers upriver from the 1760s.[16] In short, at its "founding" the city of Saint John already sat squarely in a larger geography of military contest, settler ambition, and a long-standing genealogy of commercial networks.

"Conservators of the Water": A City by Charter

What sets Saint John apart, at least in Atlantic Canada, are its origins and legal standing as a colonial municipality. Created by a Crown charter in 1785, Saint John claims to be Canada's first city, as the first incorporated city in what would become Canada. City charters defined "a body corporate and politic" by outlining the "powers, liberties, privileges," structure, and membership of the city government as a legal entity within the colony.[17] The Saint John charter thus gives us the opportunity to study closely how nature – and specifically coastal nature – was understood and organized as a political jurisdiction in British North America at the onset of the industrial era.

First, the charter explained the justification of awarding Saint John status *as* a city: as an acknowledgement of those Loyalists who, "by their exertions, conquered many of the difficulties attending to the settlement of a new country" (the kind of language that would serve the Loyalist mythology so well). It then defined the geographical boundaries of the city and brought together, at least in governance, Parrtown (the peninsula on the east side of the harbour) and Carleton (on the west side). But the most interesting piece of the charter – at least, to an environmental historian studying urban water – is the article that names the city government "conservator" of the harbour system, granting

> unto the said Mayor, Aldermen and Commonalty, and to their successors, that they and their successors be the conservators of the water and the river, harbour and bay, of the said city, and shall have the sole power of amending and improving the said river, bay and harbour, for the more convenient, safe and easy navigating, anchoring, riding, and fastening the shipping resorting to the said city, and for better regulating and ordering the same.

This was unusual but not entirely unfamiliar. In 1605, James I of England had issued his first charter to the city of London, which recognized that from "time out of mind," the mayor and citizenry had exercised the "office of Bailiff, and Conservation of the Water of the Thames" and so formally conferred this authority "upon whatsoever bank, and upon every shore, and upon every wharf of the same water of Thames." Another charter granted by his successor, Charles I, clarified the city was responsible for "improving" the river as needed for navigation and "observing of good Order, in taking of fish." The charters also entitled the city to relevant fees and "rewards" (tied to the "office of measurer," who was responsible for merchandise trafficked on the river into the port).[18] Saint John's charter, though, was more directly influenced by that of New York City's, granted in 1730. This named the mayor as "the bailiff and conservator of the water of the North [Hudson] and East rivers" and confirmed the city's title to "land over water" out to four hundred feet beyond the low-water mark of the tide, "together with all Rivers, Rivulets, Coves, Creeks, Ponds, Waters and Water-Courses." The New York charter thus defined the corporate estate of the city and the responsibility of the Common Council to "conserve" the value of the property. In leasing the highly desirable waterfront lots, the city could require petitioners to build the kind of infrastructure, like streets and wharves, necessary to create a major seaport – through the leverage of market forces.[19]

But unlike for London or even New York, the charter of Saint John was expected to *create* a city, to literally lay the groundwork from the outset. As such, it is a revealing statement of how colonial elites and British authorities saw the potential of a "new" environment. The charter draws on British legal traditions and contemporary practices of empire, but it also reveals settler expectation, anticipation, and ambition. As at New York, the goal was to create a seaport by "conferring on the Corporation the several powers and privileges usually granted to mercantile towns of the encouragement of commerce." Although the city boundaries were relatively conservative – encompassing the Parrtown and Carleton peninsulas and only to the low-water mark along the "straight shore" of Portland – the charter catalogued the environment *within* those boundaries in expansive ways. It awarded to the city "all other lands or ground whatsoever, covered or uncovered with water" within the city limits (with the exception of the "King's Land" below Fort Howe), plus islands near the falls

> together also with all and singular the woods, underwoods, trees, timber, seedings, pastures, meadows, marshes, swamps, ponds, pools, rivers, rivulets,

> fens and streams of water, land covered with water, bays, inlets, harbours, fishing, fowling, hunting, hawking, mines, minerals (gold and silver mines only excepted) and all other profits, privileges, advantages, emoluments, hereditaments and appurtenances whatsoever, to the said lands and premises, within the lines, limits, and boundaries of the said city of Saint John.

The charter thus bounded the city estate, catalogued ecological features but also human profits (fowling, hunting), and fixed the dynamic and changeable nature of the tidal estuary by itemizing its component parts. It was a veritable little kingdom.[20]

That the city was given responsibility for "amending and improving" the harbour and waterways is also telling. The concept of "improvement" permeated European colonization of North America, centring the need to order and elevate nature into a productive and appealing state.[21] That the existing landscape at the mouth of the Saint John River – an ice-free harbour, an enormous and fertile river valley, and unmatched forest stands – was deficient without human intervention to render resources into capital assets demonstrates the expectations for the purpose and utility of nature and the intentions of commercial growth that framed the colonial era. Improvement was the physical expression of human intervention for material benefit. Land that was not yet arranged for profitable purposes was deemed "waste" land until "reclaimed." (Such language would pervade North American bureaucracies of land management through the twentieth century.) Improvement was a civic mandate – a shouldering of the biblical obligation of stewardship, an application of reason, and a path to capital growth.

This last was key. The charter of Saint John – like those of New York and London – made it clear that water and shoreline would be the essential source of revenue for the city. The city's financial viability depended on its capacity to "amend or improve" the water to make it as productive as possible in a transatlantic economy. This meant a constant negotiation between private interests and the harbour as commons. The city could lease water lots and charge fees for anchorage, wharfage, and dockage. In a legal finders-keepers, it could collect any "treasure-trove, deodands, and wreck of the sea" found on the shoreline, and it could prohibit residents from raiding the beach for building supplies.[22] Council responsibilities in the harbour were primarily to facilitate commercial shipping (though without infringing on private property), whether in constructing a lighthouse and other aids to navigation or cityside wharves, piers, and storehouses. "Conservator" in this tradition reads as a trustee of financial resources more than legal guardian.

Port cities from the early modern Atlantic forward formed a kind of urban archipelago, islands of shorefront development linked by networks of extraction and exchange. They served as entrepôts that both drew and directed flows of people, materials, capital, and ideas. Their success rested on their ability to attract ocean-going commercial traffic, on the one hand, and their access to an extractive hinterland, on the other: put simply, having something to buy and something to sell. Thanks to its location, Saint John had plenty to sell to the imperial Atlantic, as one contemporary observed:

> St. [sic] John's river opens a vast extent of fine country, and takes in its various courses a number of fine rivers; on all which are rich meadow and intervale lands. And most of them settled and under improvement. The upland is in general clothed with timber trees, such as pine and spruce, hemlock and hard wood, principally beech, birch, maples, and some ash. The pines on this river are the largest to be met with in British America, and afford a considerable supply of masts for the royal navy.[23]

Because of the central artery of the Saint John River, the value of timber stands, and the Fundy tides, Saint John exercised influence over a larger hinterland than the other Atlantic ports – not only into New Brunswick but around the Bay of Fundy. The city was early a nucleus for commercial exchange, especially pronounced here because unlike its peers (or competitors) it was *not* also a provincial capital with legislative, cultural, or educational institutions. As historian C.M. Wallace wrote, "Commerce, industry, capacity and capabilities were the features paraded before visitors, not institutions, conventions and scenery ... Whatever else may be said of Saint John, it was an entrepôt with metropolitan aspirations."[24] A poem published in *The Royal Gazette and the New-Brunswick Advertiser* in January 1795 asked rhetorically,

> What raised this City, on a dreary coast,
> Alternately presenting Rocks and Frost, ...
> 'Twas commerce – commerce smoothed the rugged strand.
> Her streets and buildings overspread the land.[25]

All of this was possible because of the environmental conditions, especially of water, fresh and salt: the course of the Saint John River, which funneled lumber to the harbour; and the energetic rise and fall of the tides, which kept the port ice-free in winter and left flats exposed for beaching

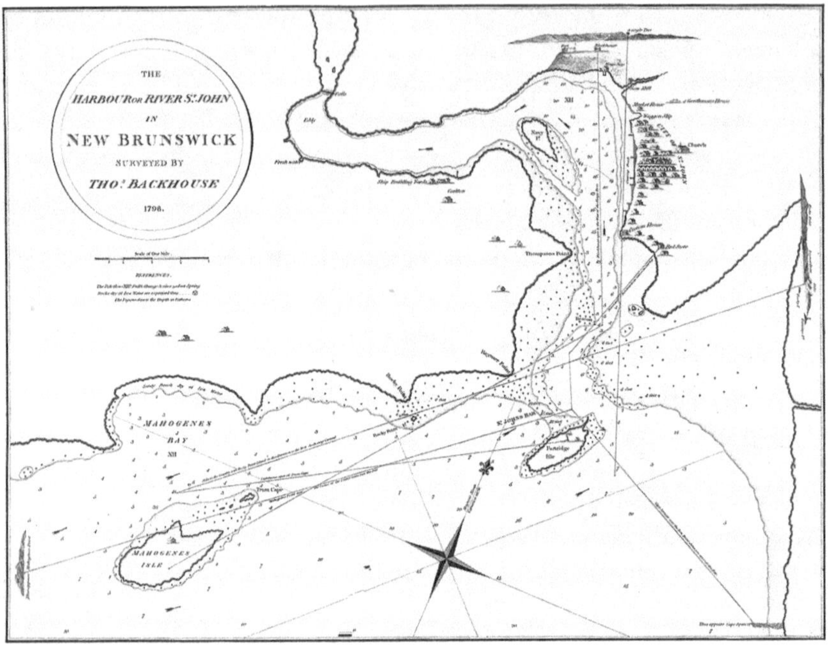

"The Harbour or River St. John in New Brunswick Surveyed by Thos. Backhouse" (1798). | The Huntington Library, San Marino, California, 105:527 M

wooden-hulled ships. The 1798 survey by Thomas Backhouse (above) seems oddly imprecise in its shore detail, but it is clearly more interested in the commercial ventures underway in shipping and shipbuilding. It shows the market and custom house, shipyards and a graving beach for hull maintenance, a sawmill, and the Market Street slip that would be the original docking point. Indeed, it forms a nascent "mercantile triangle," with commercial activity from sawmill to storehouse claiming as much of the shoreline as possible and the church set back and above.[26] Despite the image of noble privation that coloured the Loyalist mythology, by the end of the eighteenth century the city's commercial aspirations were apparent and the engines of its economic future – lumbering and shipbuilding – already established. But this activity was still shaped by the existing shoreline, for better and for worse. As Moses Perley recalled a half-century later, it was not always a shoreline of convenience:

> All the carting and carriage between the upper and lower part of this City, was carried on by the beach at low water – along by Pettingell's Yard, and Reed's Point. This way was very rough & difficult ... the beach was

> encumbered with large masses of rock. Dock Street was for a long time, only a narrow foot path, along the edge of a rocky cliff ... At low water, they went over to Portland by crossing on the flats from York Point to the Point to where Messrs. Harris & Allan's Foundry now stands. It was three years before a Bridge was got across. There was no road leading out of Town for some time – the only outlet was by a path along the bank of the Marsh Creek.[27]

The peninsula's shoreline had not yet become the town's waterfront.

Improvements and Encumbrances: Concerns of an Emerging Port

Though written as a blueprint for an urban future, the city charter was, like any document, also an expression of past experience and contemporary concerns. The expected "improvements" were those desired in the age of sail: providing infrastructure, such as docks for "fastening the shipping" and one of the earliest light-stations on the Atlantic seaboard, and attempting to regulate activities that resulted in obstacles and "incumbrances" to navigating the harbour and lower river. A particular and perennial concern for city officials was the dumping of ballast. This was the stone and rock or other filler that ships carried in their hulls for weight and balance before dumping it in the harbour and replacing it with timber for the return voyage.[28] But the city also struggled with another kind of balance: the balance between awarding private ownership, in the rush for prime waterfront leases, and protecting public access.[29]

This was partly due to the fact that Saint John was really a composite of both estuarine ecology and city organization. Acheson called the city a "little commonwealth."[30] Despite the façade of unifying frameworks – the charter, a grid on both the east and west sides of the harbour – there were significant differences between the Saint John, Carleton, and Portland shores. Each had a particular relationship to the tidal estuary: Carleton had the widest flats, Saint John the densest and most sheltered harbour, and Portland (much of which fell outside the city boundaries) was closest to the falls and upriver timber stands. The commercial core on the Saint John peninsula with its leased wharves ran up against garrison lands and breastworks at Lower Cove held by the British government.[31] Water, and wetlands, often demarcated these divisions and was itself segmented based on utility into mill ponds or timber ponds. This was not, in short, a natural civic commons.

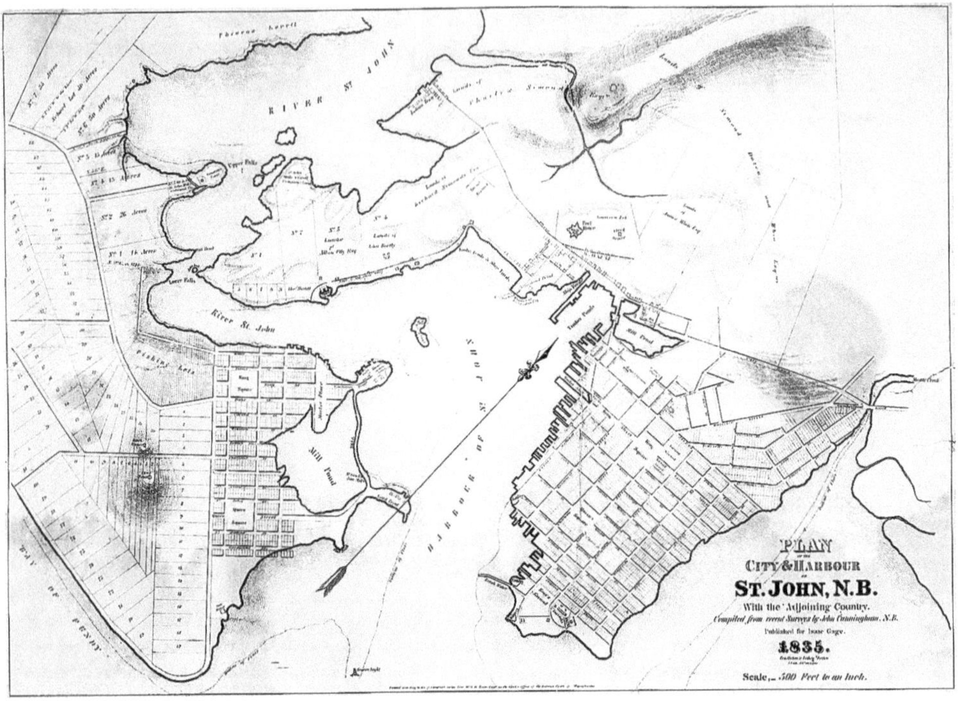

John Cunningham, *Plan of the City & Harbour of St. John, N.B. with the Adjoining Country* (1835). | Provincial Archives of New Brunswick, H1-203.60-1835/ Map MC3326-2962

Saint John's fortunes rose with British demand for New World timber and the flow of emigrants after the end of the Napoleonic Wars. Merchants embraced the charter's instructions to supply piers and wharves, making a wholly new outline for Water Street. The older beach – which would have been, at high tide, "land covered with water" – was now, as the 1835 map above shows, a geometric array of parallel wharves and slips: a breakwater was an effort to steady the waves in Lower Cove for smaller boats, and an array of dykes and dams enclosed holding ponds for timber for nearby sawmills. Given how much the city relied on revenue from leasing corporation shoreline and harbour fees, and its inability to finance deepwater wharves independently, it had some incentive to crowd in as many occupants as possible. As in St. John's, Newfoundland, this ironically became an impediment to navigation, as merchant interests built out beyond the low-water mark. Accordingly, Abraham Gesner admitted by 1847, "The harbour of Saint John is safe, but not very spacious."[32]

The contour of the harbour changed further with the construction of a canal (also visible in the Cunningham plan) at the Upper Falls. Northeastern North America was wildly enthusiastic about canals in the 1820s: Nova Scotia had begun a project to connect Halifax Harbour with the Bay of Fundy along the Shubenacadie river system in 1826, and there were already proposals for a canal to cross the Chignecto isthmus, linking Fundy with the Gulf of St. Lawrence.[33] But the Saint John canal, incorporated in 1834, was less about regional transportation than supplying local energy to "afford good Water Power for driving Mill and other Machinery to ... materially benefit the Trade of the said City, and be of great Advantage to the lumbering Interests of the Province."[34] This was an opportunity to exploit a particular corner ("neck") of the harbour and turn a navigational headache (the river current, seasonal flooding, and the vertical drop of the falls) into a profit.

By the 1830s, though, the curve of growth was confronting troubles of its own making. The city petitioned the colony's lieutenant governor for an investigation into the health of the harbour, explaining that despite its best efforts *as* "conservator," making "divers wholesome Laws and Ordinances for the preservation and protection of the said Harbour," councillors were concerned that the "yearly increase of soil" into the harbour would narrow the anchorage "and incalculable damage would be done."[35] Where was this soil coming from? There was certainly some suspicion that it included runoff from upriver, which, a former mayor testified solemnly, would only "increase exactly in proportion as the wilderness lands decrease, and the plough is used."[36] Some blamed the practice of extending wharves too far out into the harbour, which broke the current up into "eddies"; trapped, swirling, between the wharves, these dropped silt in the ever-shallower inshore waters. But more frequently witnesses pointed to new sources of waste building up the harbour floor: gravel runoff from poorly built wharves and city streets, and above all, waste from the timber and shipbuilding yards: "The vast quantity of soil washed into the Harbour from changing the Portland Mill Creek – saw dust – ballast from ships – slabs – sunken timber – chips from timber and ship yards, and various other causes." The report quietly ignored these causes and the concerns about timber waste and concentrated on regulating wharf construction. The colonial assembly was happy to pass its benign recommendations into law.[37]

Even while some residents were worried about the "increase of soil ... made to the shores of the said Harbour," others were working diligently to fill *in* the shores of said harbour and surrounding wetland. Much of this activity was focused on building out from the commercial downtown,

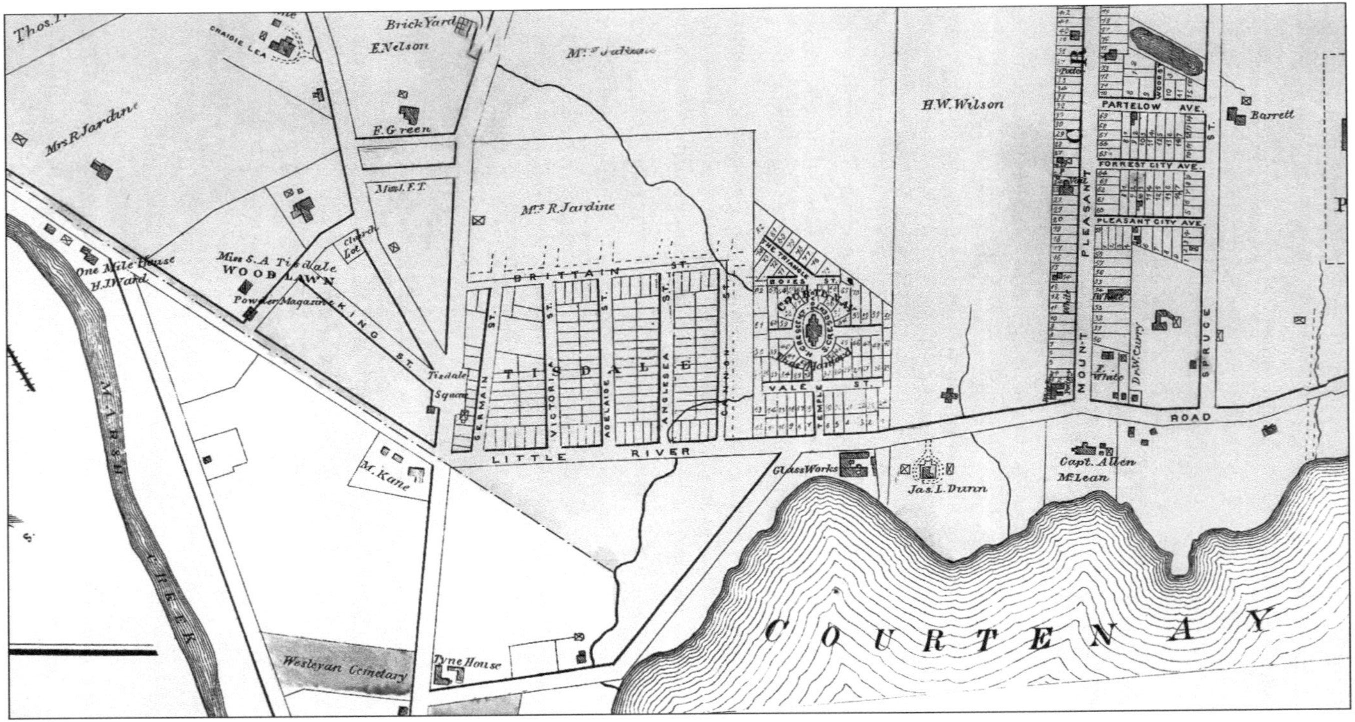

F.B. Roe and N. Geo. Colby, "Part of Simonds including Crouchville, Courtenay &c." in *Atlas Of Saint John City And County, New Brunswick. Compiled and Drawn From Official Plans and Actual Surveys* (Saint John, 1875) (detail). | David Rumsey Map Collection, David Rumsey Map Center, Stanford Libraries filling in both the "land covered by water" and the older Dock Street with the eponymous claim of Saint John Street.[38] It also moved landward, in the tradition of dykes and *aboideaux,* the system of drainage built by Acadians across the Maritime region from the mid-seventeenth century. An early concern was maintaining a bridge across Marsh Creek for public transportation and – as was generally the case with urban wetlands – for shutting out tidewater in order to drain the marsh into solid land.[39] But while the Acadian practice expanded agricultural and pasture lands, here draining wetland served another purpose. In 1825, Peter Fisher lamented the lack of attention paid to "ornamenting the City" (which he blamed on the presence of merchants with "no local attachment to the soil") but noted hopefully that "a few seats have, however, lately been begun on the Marsh near the City, which will soon make an alteration in the appearance of the Suburbs."[40] The 1875 plan on page 36 shows the kind of tidy and respectable suburbs that would have met with Fisher's approval, now laid out above Courtenay Bay. But there was little aesthetic relationship between drainage for such residential neighbourhoods and infill along the city's waterfront. The mill ponds in Portland and then Carleton would be filled in for rail and shipping, thereby cementing, literally, the economic identities of each part of the city. By the turn of the century Courtenay Bay, too, would be solidly industrial.

It appeared that the degraded state of the harbour even by the 1830s was merely the cost of doing business – business that, it had to be said, was going rather well. Despite the familiar turbulence of nineteenth-century urban life – sectarian violence, fires, outbreaks of disease – Saint John wanted the world to see a "Liverpool of America"; a city poised, as the lieutenant governor promised, to join "the first ranks of the Commercial Cities of this Continent." This was certainly the image Peter Fisher presented in 1838:

> This is the most important and wealthy town in New-Brunswick, and may well be called the New York of the Province. The trade of this City, considering its age, is immense, and rapidly increasing, and from the excellence of its harbour, accessible at all seasons of the year, and the enterprise of its merchants, it will soon attain and hold a high rank among the first trading cities of America. It is now the emporium of New-Brunswick.[41]

All that was needed, as Lieutenant Governor Harvey said, was *just a bit more development.*

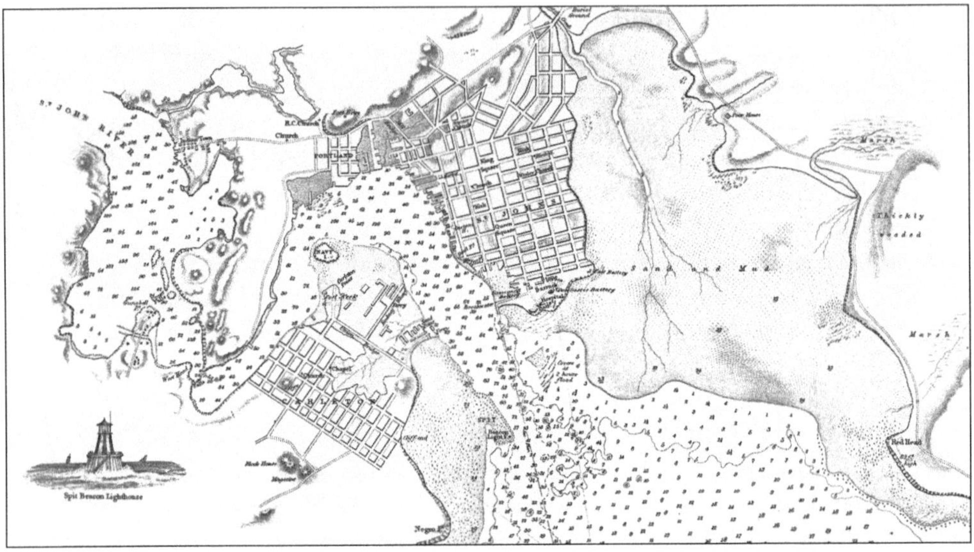

New Brunswick, the Harbour of St. John, surveyed by Lieuts. Harding and Kortright under the orders of Captain W.F.W. Owen R.N., 1844 (London, 1848) (detail). | McGill Library, Rare Books and Special Collections, rbsc_map_bayfield-128-19305, archive.org

Shore Marks of Modernity

Admittedly, much of that development, and the most dramatic changes, would come after 1880, with the expansion of railway lands and wharves in Carleton and the filling of Courtenay Bay. At first glance, the survey above from the Royal Navy at mid-century would suggest that Saint John had not yet hit its industrial stride.[42] Apart from the outline of wharves in the downtown core and a few tentative ventures onto the Carleton flats, the shape of the harbour – and more to the point, the natural features of tidal flat, marsh, and fresh water – is recognizable in the same contours as Robert Bruce had found it eighty-five years before. The Admiralty's primary concern here was still charting a safe passage through the tidal shallows. But if we look for signs of the emergent Anthropocene, such as the growth of industrial capitalism, expansion into wider networks of extraction, and an ethos of growth predicated on fossil fuels, "rapidly increasing in ... extent," we can see all of these at work in the first half of the nineteenth century.

As early as 1812 the colonial legislature issued licences for "a Passage Boat to be worked by Steam" carrying passengers and freight between Saint John and Fredericton, a much-desired development given the downward

George Neilson Smith, *View of the City of St. John, New Brunswick from Sandpoint, Carleton* (1848). Today, the cove below Reed's Point is entirely filled in, and predicted to be soon covered by rising sea levels. | Bibliothèque Nationale de France, i.d.ark:/12148/btv1b10101850c

force and current of the Saint John River, despite the perennial dangers of the new technology.[43] But such local deliveries were quickly overshadowed by seaboard and transoceanic traffic. These required larger slips and dockage, which were laid out first at Reed's Point, at the southern tip of the Parrtown peninsula, and then on the Carleton flats. The rationale was, tellingly, that inadequate "accommodation of Sea-going Steamers ... is highly injurious to the Trade of the City of Saint John"; that is, the beach and flats as found were an inconvenience, and indeed, an impairment that needed to be addressed – *improved* – for the city's well-being.[44] The bucolic foreground in George Smith's *View of the City of St. John* – a typical convention of urban portraits of the eighteenth and early nineteenth centuries – contrasts with the stages of harbour development around the mouth of the Saint John River. Logs float in the mill ponds sectioned off on the Carleton shore (to the left), while across the harbour, the congestion of spars at Market Slip marks the commercial centre. To the right, we see new wharves, and a breakwater building out into the cove south from Reed's Point, beginning to frame the cove as buildable space and a site for capital investment.

By the early 1830s, the Montreal *Gazette* touted "communication by steam extending all along the coast of British North America," including up and down the Saint John River. Readers would have found this congratulatory note printed directly below a note about fires in the coal mines at Pictou that had been burning all that winter.[45] The proximity of mines in Nova Scotia's Cumberland County would certainly fuel Saint John, despite initial hopes for productive coal fields closer by. Peter Fisher and Abraham Gesner both wrote enthusiastically about veins near Grand Lake, about 140 kilometres upriver from Saint John. As Gesner put it, "It only requires the application of capital, and enterprise, to render it subservient to the wants of the country, when such conditions of its existence have been ascertained."[46] The application of capital and technology to the extraction of fossil fuels, to render nature "subservient to the wants of the country" – there may be no better encapsulation of the origins of our present environmental dilemmas, or even of the Anthropocene.

Though coal production was never as important for New Brunswick as Nova Scotia, its availability reshaped the urban shoreline at Saint John. The use of steam power accelerated dramatically in the city's milling sector in the 1840s – and so did the effects.[47] Not even the current of the Saint John River could hope to carry off the masses of dust and edgings thrown into the harbour. By mid-century, sawdust lay five feet thick along the Straight Shore at Portland. It was enough of a concern that in 1849 two of the city's leading members of the merchant community chaired an investigation into the sawdust situation. John Murray, a local pilot, testified:

> The constant outpouring of Sawdust, edging, and rubbish from the Saw Mills in and near the Harbour, has materially injured the anchorage, and has been for late years filling the Harbour to an alarming extent; and I feel it is not too much to predict, that should the evil continue unchecked for ten years longer, the Harbour of Saint John must be almost wholly choked up.

Fishermen reported catching fish with their mouths full of sawdust. But the only alarm raised by the report was for the health of the shipping trade; not for the health of the fishery and certainly not for any larger sense of ecological well-being. As the authors warned, "The tendency of those evils, if long permitted to continue, is to produce a most serious injury (if not a future destruction) to the Harbour of Saint John, and

H.F. Walling and W.E. & A.A. Baker, *Topographical map of the counties of St. John and Kings, New Brunswick* (1862) (detail). | Norman B. Leventhal Map and Education Center at the Boston Public Library

therefore to the prosperity of this Port and City."[48] Unsurprisingly, such bureaucratic scolding and correspondingly lukewarm legislation did not seem to make much of a difference.[49] But the report's triangular equation, of harbour with port and city, made it clear that geographical features, civic identity, and economic "prosperity" were tightly aligned.

In 1861, the mayor of Saint John, Thomas McAvity, commissioned yet another report on the state of the harbour from engineer Thomas Keefer. Unlike his predecessors, though, Keefer decided to measure the current state of affairs against historical documents:

> I obtained an old hydrographic survey of the harbor made in 1761, one hundred years ago, which bears evidence of having been carefully executed. As it appears that until very recently the harbor was exposed to damage by the refuse of saw mills and the want of sufficient precaution in the discharge of ballast, it [the 1761 plan] would be of little value, even if it were possible to attempt to establish the ratio of encroachment for any given period, because this has been made under circumstances which no longer exist.[50]

That plan would have been Robert Bruce's *PLAN of the HARBOUR and Part of the RIVER S.T JOHNS in the BAY of FUNDY*, which now showed a harbour of another era. In his report, Keefer chose to focus on ballast, which had accumulated so much it had "shoaled" the outer harbour nearly twenty feet. Ballast rock was augmented by the dirt and gravel washed from the wharves and city streets, as well as material excavated from the city's bedrock to install the kind of municipal services – gas, water, and sewer – that an aspiring metropolis had come to expect. Keefer treated sawmill waste almost as an afterthought; another departure from the 1849 report but politically a savvier approach. In light of "the growth of the commerce of the City," he felt, the silted and congested harbour was not "an unmixed evil."

Keefer recommended, as had others, extensive dredging between the wharves but – evidently making lemonade out of a century of waste – also for the city to embrace a deliberate project of infill on both sides of the harbour. "The object then should be to enclose the largest area with the least proportional expense," he wrote, proposing a dramatic expansion from the breakwater southward and, when the railway reached Carleton, beyond Sand Point. As he pointed out, city plans, such as the 1862 profile on page 41, had already overlaid proposed streets and squares not just on the Carleton mill pond but outward into the harbour alongside the deep railway slips, "evidently with the intention of converting them sooner or later into dry land." The proposed extensions would dramatically expand buildable property and usable water frontage of this side of the city. Compared to older plans, the map by Walling and Baker fades the flats and then emphasizes commercial activity by superimposing a city directory on top of the city. After all, Keefer wrote, "trade will increase and require all the facilities of which the harbor is capable; and as this is the main dependence of the City and the Province, it is clearly your policy to foster it."[51]

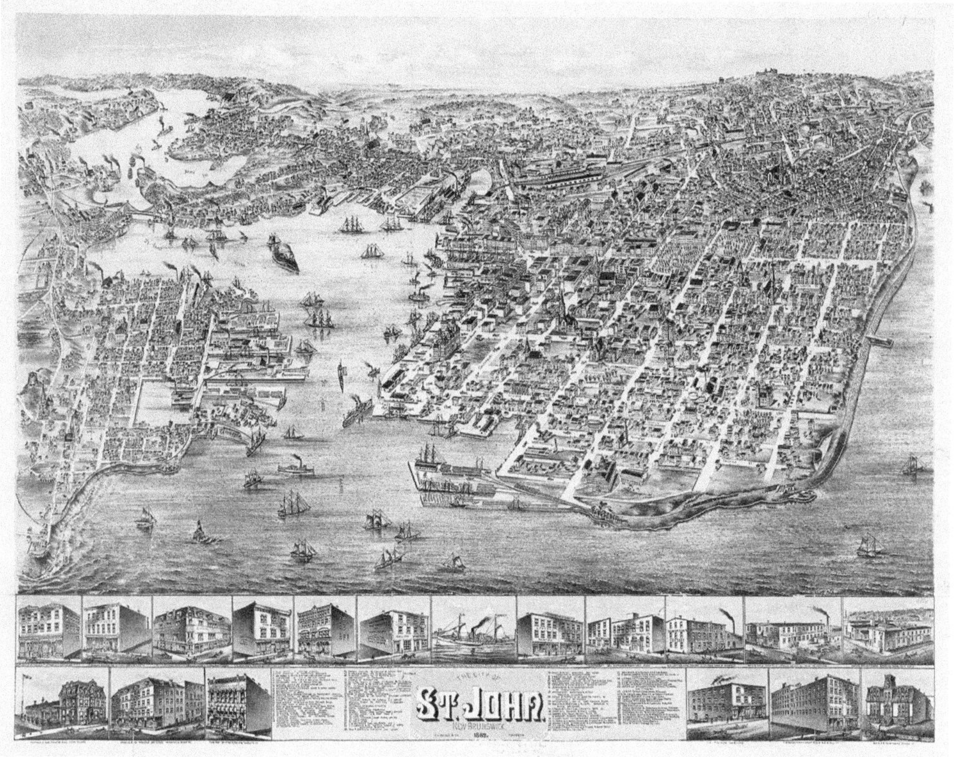

O.H. Bailey & Co., *The city of St. John* (Boston, 1882). | Norman B. Leventhal Map and Education Center at the Boston Public Library

Postscript

Saint John would indeed continue to develop "all the facilities of which the harbor is capable." Consider the dynamic scene of the bird's eye map above, drawn nearly twenty years after Keefer's report and fifteen years after New Brunswick joined Confederation. Ships under sail and steam manoeuvre past each other through a new narrows formed by wharves extending from Carleton and Reed's Point. On the Parrtown peninsula, the Lower Cove is half its original size, and there is no indication of tidal marsh or flats, only railway embankments. Smokestacks ring the inner harbour, enormous piles of lumber lie stacked on the wharves, and the suspension bridge allows cart traffic over the falls. But the most dramatic change is the overlay of the three railways reaching into the city: along the Portland valley, around to the old ballast wharf, and from the west, to the Carleton wharves.

The city had been caught up in the prevalent enthusiasm for railroads since the 1840s, "desperately seeking a place in a continental system of transport nodes in which the port could become the axis between Atlantic water and Canadian soil."[52] Foundries along the shores of the old mill pond – bordered by rather misleadingly named streets like "Arcadia" and "Paradise Row" – churned out rolling stock for railways as the city sought to secure the transport of St. Lawrence manufactures and the burgeoning export of western grain. The European & North American Railroad reached from Saint John to Shediac and the Gulf of St. Lawrence by 1860; its Western Extension (shown here as the St. John & Maine Railroad) was built to the Penobscot River a decade later. These were consolidated into the Intercolonial and Canadian Pacific lines, bringing the city into new, externally driven frameworks of federal authority and capital investment. The Canadian Pacific Railway (CPR) colonized much of the Carleton shore with loading docks and grain handling. "Shipping" in Saint John had become synonymous not only with shipbuilding but with the entrepôt functions of commodity exchange.[53] Halifax never conceded the title, but Saint John entered the twentieth century announcing itself as Canada's winter port.[54]

As the harbour filled in, as transportation and industry required more space and more distance from the urban core, the functions of the port detached from the waterfront of the city. As Martyn Bowden observed of Portland, Maine, the arrival of railroads around and at the edges of the older mercantile centre created a new geography of industrial districts on the city's outskirts.[55] By the early twentieth century the major port activity had turned to the wide tidal flats of Courtenay Bay on the eastern edge of the city, barely visible on the bird's eye map of 1882. Now the traditions of shipbuilding and export could look to a new frontier of shoreline with dreams of expansion unconstrained by the shape of the older harbour and the force of the Saint John River. Courtenay Bay would become home to the largest dry dock in Canada in the Saint John Shipyard, a critical location for wartime fleets. But it was the Irving complex that would most dramatically remake the city's outer shoreline, dredging a harbour twice as big as the original, with the first deepwater terminal on the seaboard and then the largest oil refinery in Canada, berthing tankers of unprecedented size carrying vast amounts of crude. This was the industrial suburb realizing most extensively and most permanently the city's "large conceptions" and the very arc of the Anthropocene. Soon thereafter another kind of petroleum landscape reshaped the harbour itself, with Navy Island reduced to nothing more than a bridge footing for the TransCanada Highway.

Today we are further removed from Thomas Keefer than Keefer was from Robert Bruce. The capital fixed in concrete around and across the city harbour has created a third-generation harbour that neither man would recognize. Yet Bruce's plan *is* of value to us. Without it, we would have little sense of the original shoreline, the extent of the harbour's transformation in the industrial era – or of our own responsibility for that transformation. Thus, in 1935, journalist Ian Sclanders could write exultantly, "It was natural that a city should grow, for here was a harbor and here were the crossroads of the trails of trade ... Here a great port was built. It was natural that a port should have been built here."[56] But the shape of the city was not natural or inevitable; it was the result of a mandate to "amend and improve" the waters and a prevailing ethos of commerce and expansion. Nature offered an invitation, and the city accepted with enthusiasm.

2
Limits Defied
St. John's, Newfoundland

Situated on the eastern edge of the Avalon Peninsula, on the southeastern reach of the island of Newfoundland, St. John's is a city of the North Atlantic. It has served as a port and staging ground for seagoing traffic since the sixteenth century, from fishing fleets to wartime convoys, and now offshore oil supply ships. St. John's has long been considered a refuge against a forbidding northern climate and stormy North Atlantic seas: a tight turn through the Narrows (only three hundred feet wide) opens into a compact harbour angled to the southwest, sheltered by the steep and imposing hills of the South Side. Climbing the wind-buffeted paths of Signal Hill above the Narrows affords a view of the harbour that appears to have changed little from centuries past. Visitors have long found the encircled harbour charming, a kind of geological amphitheatre, a "mighty Coliseum ... with a little city looking down from the mountain side, and a swamp of shipping along its shores."[1]

A little city, indeed: St. John's grew slowly compared to other seaboard cities. Despite the geographical advantages of its location for sailors, on the doorstep of the New World, the city has a Sisyphean history of development: spurts of growth in population and urban services frequently reversed by disaster and turns of economic fortune, confined by season and weather, and constrained by a distinctive and tendentious character of colonial government and community. St. John's may be founded on a rock, but the city sat precariously on its shore.

Yet the faith in growth – in urban expansion and permanence – that characterizes the modern city is apparent in St. John's throughout the long

William Eager, *The Town and Harbour of St. John's, taken from Signal Hill, June 1st, 1831.* | National Gallery of Canada, 30240

nineteenth century and well in place by the dawn of the twentieth. The limits of space and season within the harbour's walls were more than matched by hopes for, and dependency on, the ocean beyond. It was, presumably, there for the taking. As a blue frontier, the ocean represented a vast storehouse of wealth to dazzle the imagination and that could legitimately be brought within the city's ambit as a hinterland of harvest. As one visitor remarked in 1881, "They have a saying that an acre of the sea is worth a thousand acres of land."[2] The St. John's waterfront was where the wealth of the sea was brought ashore. Whether for the Grand Banks fishery or oil on the Scotian Slope, the architecture of the waterfront has expanded in pursuit. Earlier interventions were small, but they laid the groundwork – literally – for more ambitious, irrevocable transformations in the industrial era amid the turn to fossil fuels.

This, then, is the story of urban change within a constellation of constraints: spatial, seasonal, climatic, geological. It is the story of how a coastal city sought to defy environmental limits by using "made ground" to create useable space for resource extraction. But it is also the story of the tensions and costs of constructing a shore. By the 1880s, these tensions were articulated in extensive debates over "encroachment," shore property, and the purpose and pliability of the harbour. St. John's has the longest history of harbour interventions of any of the cities considered in this book; archaeologists suggest the water area has been decreased by approximately a third since the earliest European arrivals in the sixteenth century.[3] While the colourful houses of Jellybean Row and the brick street fronts of Water Street appeal to the tourist as glimpses of a historic city, the real story is how much, how far, and how many times the city's waterfront has been rebuilt. And to what end.

Early St. John's

Unlike the other cities featured in this book, St. John's has no clear or known founding date: it was a city born of utility and practice more than intention or plan. The harbour is the site of one of the longest histories of European occupation in the New World. Even as Jacques Cartier explored the Gulf of St. Lawrence in the 1530s, migratory fishers from Brittany and Normandy were landing on the shores of Newfoundland for a few weeks in the summer. England claimed the island it named Newfoundland as a colony in 1583, likely more as a fishery hinterland than a place for settlement, though small "plantations" were granted by the early seventeenth century.[4] Newfoundland's fishing banks and landing coves became an increasing point of contest between France and England through the next hundred years. The Treaty of Utrecht awarded the island to Britain but gave France fishing rights to the northeast "French Shore," leaving Newfoundland in the crossfire of the ongoing conflict between the two empires during the eighteenth century. The British erected small bastions at Signal Hill and the mouth of the Narrows, along with Fort William above the harbour, but it was not until the later part of the century that Britain began to construct more robust naval facilities and fortifications in the town.[5] Visitors praised the "safe capacious Harbour ... sheltere'd from all winds by great and stupendous Hills," though they were aghast at the reach of winter weather. As Aaron Thomas noted in 1794, "In this Empire of Frost and Snow great quantitys [sic] of wood must be used by the Inhabitants whose winter here is from October to April! – or May!"[6]

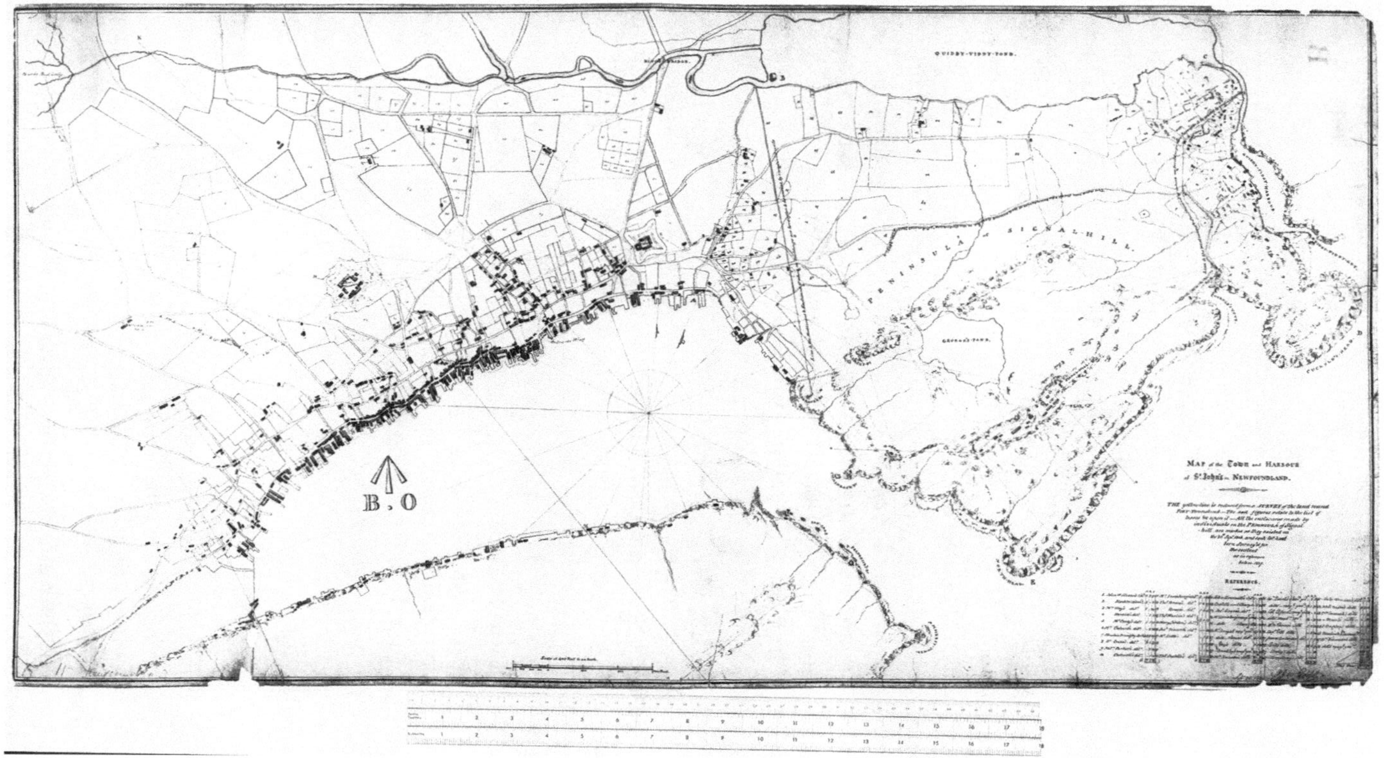

J.G.W. Eastaff, "Map of the town and harbour of St. John's in Newfoundland" (1807). | Library and Archives Canada, 79001/64, local class no: H2/140/St. John's/1807

The steep tilt of the harbour's south side, and the utility of the shore relative to "the barrens" along the hilltops, confined the built landscape of the town along the northern edge of the waterfront for centuries. Two streets, known as the upper and lower streets, ran along the harbour, following terraces ascending to Signal Hill in the east. (That there were only these two paralleling the harbour indicates how close the town stayed to the water until the mid-nineteenth century.) The most important was the lower street, Water Street, which may be the oldest street in a European settlement in North America and which, as the name suggests, followed the curves of the waterfront. By the mid-eighteenth century it formed a thin and chaotic corridor of commercial activity, hemmed in by the steep slope of bedrock while reaching into the near waters of the harbour. Coves separated by spurs of gravel along the north shore of the harbour supplied landings for small boats while wooden flakes, or platforms for drying fish, were laid out alongside. The coves and the streams that emptied into them began to be filled even during the eighteenth century but have been preserved in street names like Beck's Cove and Ayre's Cove.[7]

For most of its history, the town's growth was "unplanned and ungraceful."[8] This was partly due to the topography but more to its raison d'être as a fishing port and the immediate prescriptions of a competitive working waterfront. Unlike chartered settlements or model towns, St. John's fell under "the rule of the admirals," a form of naval government whereby the first fishing crews to arrive in port each season claimed judicial authority over "ships' rooms," or landing beaches.[9] Lacking the security of property tenure on their waterfront leases, merchants jockeying for a toehold by the water built their wharves, storehouses, and flakes quickly and cheaply out of wood. The focus on seasonal harvest and processing (not to mention the weather) actively discouraged permanent residency and thus any kind of urban planning. In 1804, for example, the governor of Newfoundland reported that those with claims to fishing rooms were now building "dwelling houses on the ground near the sea" behind their flakes. They were not alone. As the governor wrote disapprovingly, "The north side of the Harbour is now taken up with merchants Stores Wharfs &c. for the purposes of Trade ... the sea-side is crowded with Dwelling Houses, Retail shops, and Public Houses, with various Artificers." The mercantile streetscape might be necessary, but, he argued, it was the "Flakes, and other erections for curing Fish" that should be given priority and afforded "every convenience and facility."[10]

By the end of the eighteenth century, though, consistent metropole demand for saltfish and seal products had helped plant a resident population in the town. St. John's also benefited from its status as the most secure port on the island, the naval headquarters, and home of the governor.[11] Despite the perennial uncertainty of a resource economy – with the tumults of overseas markets, weather, and so on – economic growth through the Napoleonic Wars concentrated the attention of the increasingly restless merchant class on the question of landholding. In 1811 the British government acknowledged the reality of the situation (which the former governor had resented) and permitted the formal leasing of fishing rooms as "Spaces of Ground ... let out for building Dwelling Houses and Store Houses, and for other Uses necessary to the Trade and Fishery ... [and] may be disposed of as Private Property."[12] But despite this local accommodation, land title was still generally held by absentee landlords. Governor Richard Keats observed that there was a particular problem in downtown St. John's, where those "Occupiers of Land by encroachment" needed to be managed more closely through leases and quit rents. Keats – who issued over a hundred leases on the outskirts of St. John's in hopes of stimulating local agriculture – recognized both the growth of the town in recent years and its economic precarity:

> St. John's became the Emporium of the Island in consequence of this extended war, with a population of nearly 10,000 inhabitants, seems to have grown out of its character from a Fishery to a large Commercial Town, and for a considerable time past has offered such advantages to the Farmer and Gardener as to surmount in a great degree all the restraints which Nature or the Policy of Government has laid on the Cultivation of a Soil, certainly less sterile than has generally been considered. But this character which it has latterly assumed, it is very doubtful if it will be able to support on the return of Peace.[13]

The new property regime would shape the urban character of St. John's, for better and for worse. In his 1819 *A History of the island of Newfoundland,* Lewis Anspach approved of the change to granting leaseholds, which he believed would encourage a new permanence and sophistication: building of stone, spaced for healthful circulation, and similar "improvements which would most effectually contribute to introduce cleanliness, salubrity, and even elegance, into a town hitherto remarkable for a confined and unwholesome atmosphere."[14] Such aesthetic aspirations could not, however, compete with present realities; imagined elegance was not as profitable as immediate export. So the convenience of building cheaply and densely in wood continued.

Conflagration and Encroachment

Now the jumble of wooden stages, flakes, storehouses, and wharves, laden with drying fish and vats of seal oil, could light sheets of flame across the harbour. As a representative from the Phoenix Insurance Company of London observed in 1817, the result was "a mass of insecurity and danger in respect to fire, hardly to be paralleled."[15] More than any legal or planning regime, fire would continually remake the St. John's waterfront. Though some would loom larger in the city's history – especially fires in 1817, 1846, and 1892 – a British engineer stationed in St. John's observed in 1835 that "Scarcely a year passes without a fire burning down from twenty to one hundred houses."[16]

Like other colonial cities, municipal responsibility for fire response and prevention was minimal, left instead to local organizations and brigades (and in the case of St. John's, the garrisons).[17] The onus on private initiative was a mixed blessing here: those most willing and able to act were those with the most capital at risk: the merchant class.[18] The density of flammable materials at the waterfront meant the commercial core would be most vulnerable, but the costs of protective measures were anathema to those most concerned with this season's profits. This was especially true given the added possibility of environmental disaster, whether storm or late spring ice, jeopardizing any harvest. As insurers recognized, building to stricter code would be more efficient in the long term but a difficult sell in the short, and in St. John's in particular:

> To entail upon the proprietors of property there the duty of making a positive and solid improvement which would ultimately be the cheapest to themselves, but it would be unpopular at such a moment because requiring a larger first expense, which in the impoverished circumstances of the place, cannot be readily supplied; and it would also require a much longer time to rebuild the place in this manner.[19]

The analogies to our reluctance to invest in climate adaptations are, I hope, self-evident.

Nevertheless, after every "conflagration," the colonial assembly marshalled new legislation in an effort to implement more robust standards for urban construction. Especially following the fire of June 1846, St. John's was prescribed wider streets and firebreaks; building and roofing in brick, stone, slate, or tin; expanded water supply from ponds around the town;

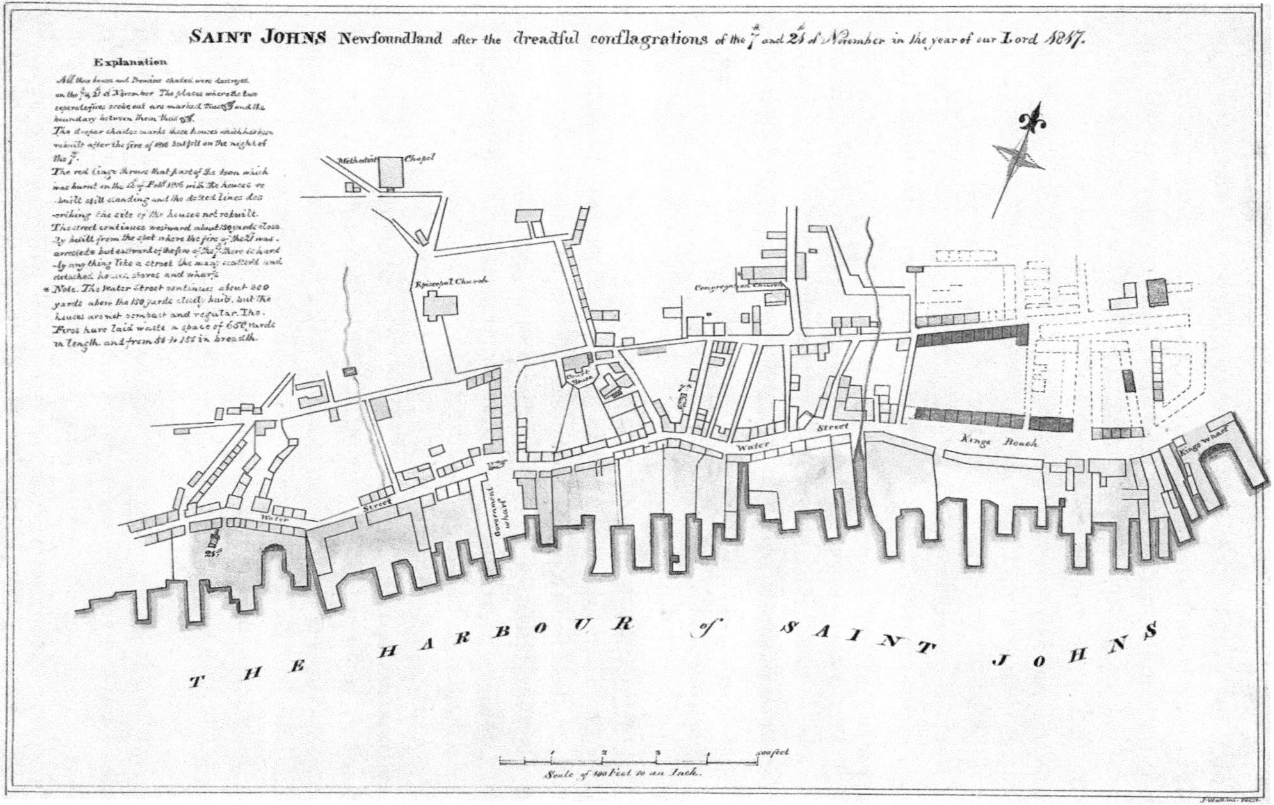

J. Walkins, "Saint Johns, Newfoundland after the dreadful conflagrations of the 7th and 21st of November in the year of Our Lord 1817." | Library and Archives Canada, H12/140/St. John's/1817

and municipal organization into wards for more effective distribution of relief and services. Some of these were carried out. More space was brought into the public domain, expanding government authority over building inspection, the spatial dimensions of the streetscape, and the removal of "public nuisances" or potential dangers to public safety. In one revealing instance, the colonial legislature instructed that the revenue from the wharf at Beck's Cove be directed toward "improving the Watercourse of a certain Stream which flows into the said Cove ... as a Tank or Reservoir for Fresh-water in the aforesaid Cove," under the supervision of the fire warden. Thus, a freshwater stream already in the process of being buried was to be partly diverted for the purposes of fire protection for the commercial waterfront – a succinct expression of mid-century municipal priorities.[20] Rebuilding was sometimes presented as an opportunity for public works and public employment, in the kind of bourgeois improvements that Anspach would have preferred. This included a new Marine Promenade laid out as a westward extension of Water Street, along the Waterford River at the west end of the harbour, in 1847.[21] As we will see, there was an even more direct throughline from the fire, especially that of 1846, to changes in the waterfront itself.

These moments of optimism, though characteristic of the age, were encircled by the island's environmental constraints and economic uncertainties. As one commentator wrote in 1841, "the ungenial climate, rocky soil, and constant recurrence of extreme distress every winter, must, at all events till some remote and uncertain period, prevent this Island from being recommended as a desirable point for emigration."[22] (Importantly, this understanding of the future and the island's potential changes radically in the era of fossil fuels.) In his 1885 history of Newfoundland, Moses Harvey characterized the first half of the nineteenth century as a chapter of "Commercial Disasters."[23] The combination of the property regime and the uneven returns of the resource economy made it difficult for St. John's to establish a base of revenue or to enforce many of the fire-prevention or planning ordinances. Though it claims to be the oldest city in North America, St. John's won municipal standing comparatively late, instead governed by the colonial legislature and funded indirectly through import duties or loans. It was not until 1888 that a town council was established to oversee services such as water, parks, and fire protection.[24] Coupled with the influence of absentee landlords – who were more interested in receiving rents from waterfront properties than contributing to city-building – it was little wonder that St. John's struggled with the costs of rebuilding from each fire.

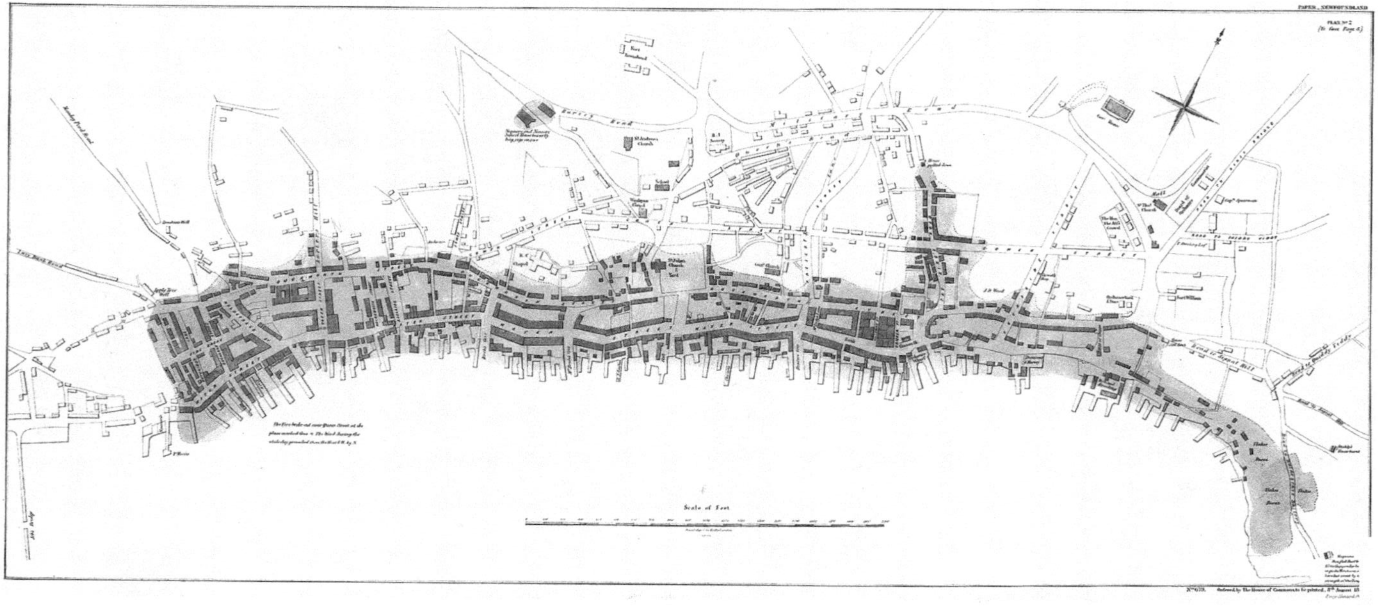

"Map of St. John's, Nfld., showing extent of fire of 1846" (1851). | Digital Archives Initiative, courtesy of the Centre for Newfoundland Studies, Memorial University Libraries

But the priority for rebuilding was always the same: returning the waterfront to capacity *and more.* Simply, hazard was the cost of doing business, with the cod fishery of the Grand Banks in summer and the seal hunt on the ice floes in winter. When visitors described St. John's as a city of one street, the geography was symbolic; the orientation and character of Water Street embodied the centrality of extraction. "The interests of the city are all with the sea," said one guidebook.[25] As Pavla Šimková has observed of Boston and its harbour islands, St. John's functioned as a city by incorporating the offshore into its urban fabric.[26] A municipal requirement to roof a warehouse in tin while storing barrels of lucrative seal oil on a packed wooden wharf was a small price to pay.

And there were other ways to enfold rebuilding into waterfront expansion. In 1997, St. John's undertook a massive twenty-year harbour cleanup project that included installing a large interceptor sewer around the harbour's edge. The excavation for the new sewer prompted substantial archaeological exploration under Water Street, which found a "fire layer," where burnt material and rubble had been built over or dumped in the harbour to make new ground. As archaeologist Peter Pope succinctly observed, "St. John's has been shaped as much by fire as by water."[27]

Nineteenth-century St. John's residents knew very well they were altering their shorelines this way. The particular shape of the St. John's harbour – with the constriction of the Narrows, unlike Halifax; with sedimentary rock, unlike Saint John; with steep slopes, unlike Charlottetown – meant there was little place to go except onto the water. Shovelling burnt rubble and roadbed rock into the shallow waters of the foreshore would have seemed a reasonable and immediate solution to clearing the streets while also creating new land in the most valuable location. Such "made ground" was a new commodity and capital asset. It was not, however, without its complications, or what environmental historians would call unintended consequences.

We have a remarkable source for this story in the harbourmaster, George Robinson, who issued a number of reports on the conditions of the harbour in the 1870s and 1880s. Robinson was a master in the Royal Navy who had been stationed at St. John's and in 1876 was appointed the first harbourmaster and examiner-in-chief of masters and mates. In 1881, the colonial secretary instructed Robinson to undertake a study of "the present condition of harbour of St. John's, with a view to the prevention of any further encroachments on its area." Robinson's report, "with various statistics and suggestions for its maintenance and improvement," is a wonderful snapshot of the traffic and transformation of the harbour to that point from someone whose concern was more maritime than mercantile.[28]

Robinson was unequivocal. It was "a matter of history," he wrote flatly, "that Water Street has attained its present breadth by encroaching on the harbour; the filling-in process took place after every conflagration." Some of the fill was burnt rubble from successive fires, and some came from blasting the north slope of Water Street in order to widen the street for firebreaks after 1846. That debris "can be plainly seen under the wharves of several of the mercantile premises." Indeed, "the whole of the waterside premises are built out of the harbour and some portion of Water Street." At the same time, Robinson noted, waterside tenants had filled in the foreshore to expand wharf acreage against the harbour slope – the act of "wharving in": "immense areas have been enclosed, in order to obtain level water-side premises, and excellent premises they are, but the old coast line of St. John's was not level by any means." In a remarkable set of calculations, Robinson measured the base area of the harbour at 305.8 acres, and the "area of enclosed water" and various foreshore additions, "built out from the ancient high water line unto this day," at 62.6 acres. Overall, he estimated, only half the harbour still offered clear anchorage.[29]

The very characterization of this expansion as "encroachment," implying a loss of space, rather than the more positive "made ground," indicates Robinson's view from the water. As harbourmaster he was tasked with managing maritime traffic in an increasingly unwieldly and congested space. To underscore (literally) his point, Robinson drew a line in red ink to mark the extent of the encroachment, as well as the wharf extensions underway. He also recorded how far each section of the waterfront had been extended since 1804: King's Beach (by now Queen's Wharf), the historic location of the Customs House, was a relatively modest 60 feet, compared to Clift's Cove, at 175 feet; Hunters Cove, 180 feet; Stewart's Cove, 215 feet. Beck's Cove has advanced 225 feet since 1804. Taken together, Robinson calculated that 38.1 acres or 1,662,679 square feet had been added to the foreshore over the course of the century.[30] Like any resource, the shoreline and the navigable waters of the harbour became more contentious the scarcer they became. But the same was true from the shore; from the merchant perspective, the solution to a crowded waterfront was to expand the waterfront.[31] If an acre of sea was worth a thousand of land, then 38 acres exactly where the wealth of the sea came ashore was a fortune.

Filling in and fire rubble were not the only issues facing the harbourmaster. Robinson had already raised concerns about the congestion of

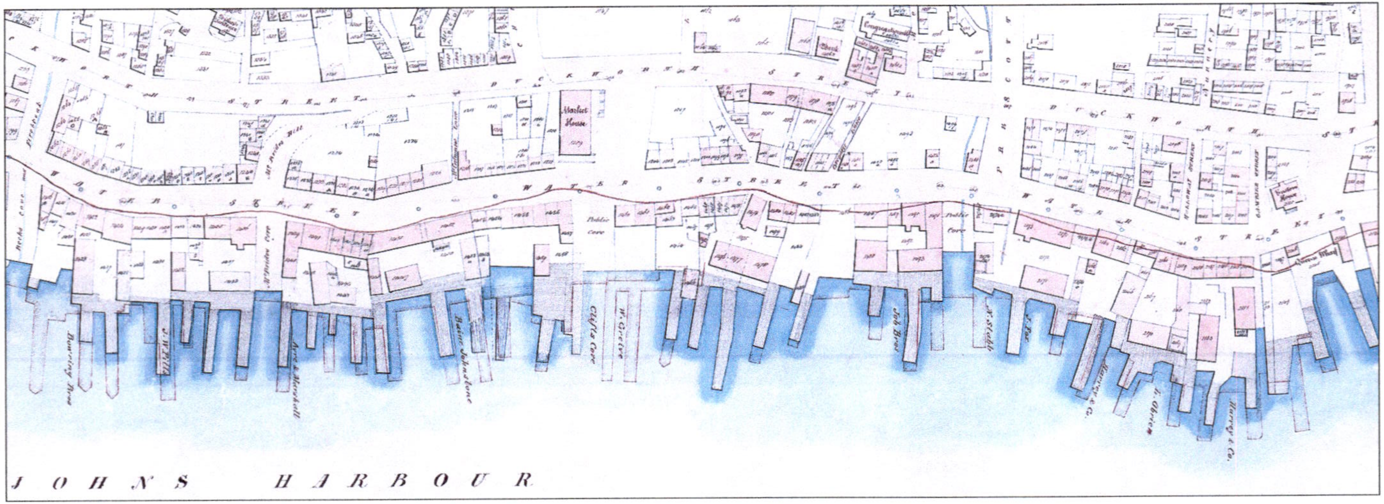

Frederick Page, "Map of St. John's, Newfoundland, shewing [showing] all the buildings erected since the fire of the 9th of June 1846 from actual survey (1849)," with 1882 addition from George Robinson, detail. | The Rooms Provincial Archives Division, MG 50.2

wharves and the accumulation of litter on the harbour bed and the effect of these on narrowing the harbour. Decades of sunken vessels, wharves, stages, and bridges, either abandoned or neglected, "have been instrumental in silting up the heretofore valuable anchorage ground." Such obstructions," he wrote, "are calculated in the course of time to effect serious injury." Chief among these obstructions were an "Old Flake," "this ancient and useless obstruction to the waters," sitting at the west end and a "mussel bank" of deposition about 130 yards long perpendicular to the wharves in the commercial core.[32] Both of these, Robinson felt, contributed to a larger problem: the constriction of water flowing from the Waterford River out through the harbour. The sheer density of wharves and wrecks slowed the stream enough to drop more of its sediment into the harbour (the same problem observed at Saint John) at the same time as the expansion of sewers in the city in the preceding twenty years had increased the silt being carried out to the harbour bed. Robinson was concerned that building not just in the harbour but into the river – the Marine Promenade being a prime example – had constricted water flow to a significant extent: "The old people tell me that the water [Waterford River] used to flow over the road at the Gas Works; it certainly does not do so now." Taken together, such a "strangulation of the tide" spelled trouble for the harbourmaster and anyone trying to navigate the harbour.[33]

As the St. John's newspaper the *Evening Mercury* pointed out, "encroachment" really took two forms: physical infill to build out the foreshore, and the extension of private wharves into the public waters of the harbour.[34] Robinson's report had been commissioned for a joint committee of the Legislative Council and the House of Assembly investigating land tenure, or the leasing arrangements between tenants and (primarily absentee) landlords. A half-century of attempts at municipal reform had faltered in the face of inertia or opposition from the landlord class, but by the 1880s the situation was acute, with a number of waterfront leases and rebuilding loans coming due.[35] The committee focused almost entirely on the most lucrative part of the city: the commercial district of Water Street fronting the harbour and now extending into it. And it felt strongly that encroachment – as in the extension of breastworks, wharves, and private lots into public waters – represented "the usurped possession of the adjoining waters of the Harbor, that are the undoubted Public domain" and required "the prompt action of the Government."[36] Tenants testifying before the committee, including some of the most prominent merchants of the city who also leased waterfront property, generally agreed that the waters of the harbour should be considered public, or "the property of the Queen."

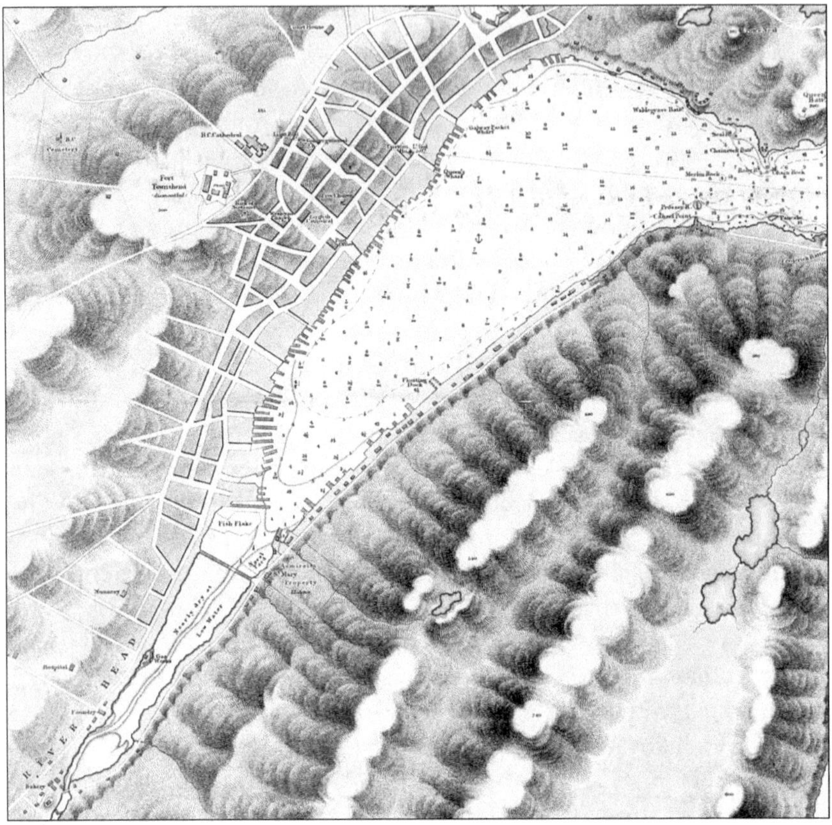

St. John's harbour surveyed by Captain Orlebar R.N., assisted by Commr. Hancock, Messrs. Carey, Des Brisay & Hyndman R.N. (London, 1862). | McGill Library, Rare Books and Special Collections, McGillLibrary-rbsc_map_bayfield-022-18902, archive.org

Indeed, a few suggested the government purchase the entire waterfront. There was far less consensus about what "prompt action" should entail and what a wharf lease permitted; some wanted the right to expand, and others did not. "I think the Merchants are the best judges," Robert Hunt told the committee repeatedly.[37]

Given this divided and contentious political backdrop, it is unsurprising that Robinson's recommendations – like those of the Saint John harbour commissions – leaned more to structural remediation than regulation. These included removing sunken wharves, wrecks, and ballast and dredging a channel, especially at the head of the harbour, in hopes of encouraging a better flow of freshwater in one direction and tidal in the other, as well as implicitly marking out space for public navigation in the harbour. Such

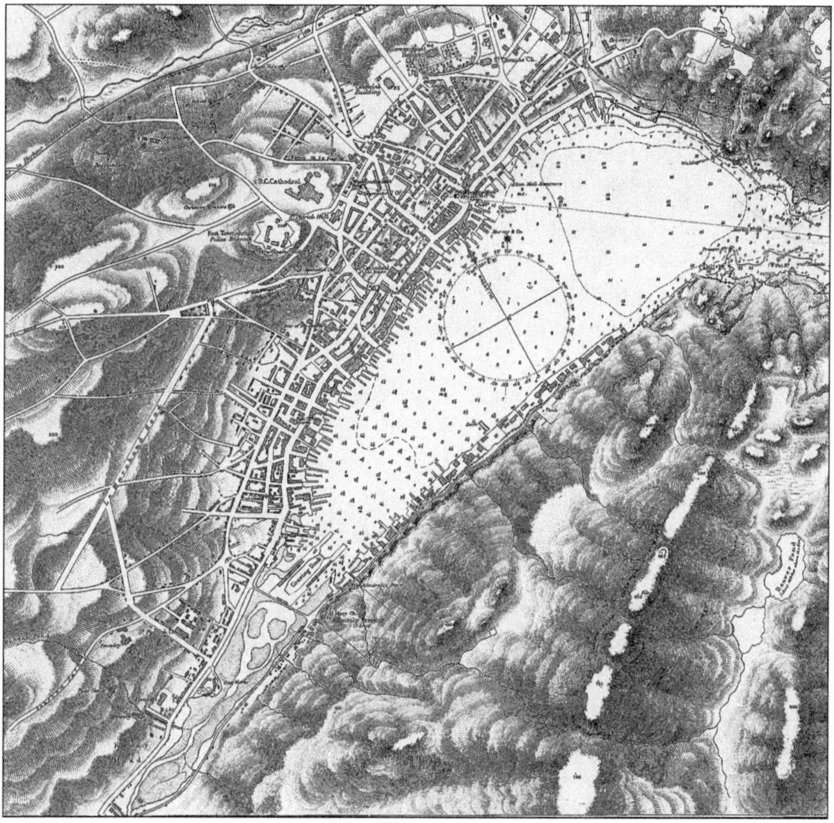

St. John's harbour surveyed by Commander G. Robinson R.N., Harbour Master of St. John's, 1885/British North America – east coast. Robinson noted that water depths had become shallower since the 1862 chart: "The old channel with six feet of water as shown in Capt. Orlebar's plan is now a dry mussel bank, the cause is plain." The other significant change was the construction of the graving dock, which walled off much of the Waterford valley and enclosed the west end of the harbour. | Library and Archives Canada, R11630-582-4-E

interventions aligned with the preferences for mechanical "improvements" and enhancing productivity; even more importantly, they were about expanding space rather than containing growth. Robinson *did* counsel a policy of thus far and no further, holding the wharves at their existing lengths (and he dutifully measured ninety-five of them to supply the government with a baseline), but that was as far as he would go in confronting the geography of private capital along the waterfront. "The only other alternative is the reduction of the wharves," Robinson concluded, "which is beyond the subject of this paper."[38]

In his landward battle against shoreline encroachment into the harbour, Robinson was prepared to allow one very large exception. A dry dock, he admitted, would require a deduction of six or seven acres of harbour acreage, but its absence was keenly felt. Domestic sealing fleets and transatlantic mail steamers alike needed a safe port of repair, which St. John's was uniquely placed to offer. Robinson shrewdly added that attracting more shipping traffic into the harbour would be good for the city's economy and further raise property values on the waterfront, but the dry dock was really about acknowledging the position of the city "in the ocean that surrounds us. It is the natural result of our situation."[39]

Expansion and Enclosure: The Industrial Suburb at Riverhead

How best to profit from that situation – "on the high road from Europe to North America," in Robinson's words – had become an important question for nineteenth-century St. John's. The transatlantic telegraph cable first laid in 1858 and regular steamer connections by the 1860s began to recast the particular value of St. John's harbour as a fulcrum for transatlantic traffic and communications, which then became a standard point in descriptions of the city in atlases and travel accounts.[40] The problem was that the Atlantic was getting more crowded, the ships larger, and their fuel (coal) was taking up more space on the waterfront, while St. John's harbour was staying the same size. While new suburbs were being laid out uphill from the harbour's north edge, most hopes for industrial growth were pinned on the west end. In 1883, Joseph Hatton and Moses Harvey imagined that the appearance of the entire city would improve with such a development:

> In due time St. John's will be transformed into a handsome city, for the magnificent site it occupies admits of the introduction of the best improvements of modern times ... Already the want of more harbour accommodation is felt, and it will soon be found necessary to deepen and extend the harbour at its western extremity, so as to admit of an increase of wharfage. The want of a graving dock is at length to be supplied ... equal to anything on this side [of] the Atlantic.[41]

So attention had already turned to the west end and the mouth of the Waterford River as the most likely site for expansion. Whereas the industrial infrastructure at Saint John's could claim purchase along the Fundy flats, no such development was possible on the exposed rocky shoreline of Newfoundland. Within the shelter of the harbour, the Narrows were too, well, narrow; the south side too steep and unmalleable; the north side too established and too valuable – which is probably why several witnesses testifying before the land tenure committee supported development at the west end. The district known as Riverhead was already established as an industrial periphery, with the city gasworks located there since 1845.[42] Moses Monroe, for example, suggested "that the Harbor might be largely increased by the enclosing of the river, as far as Gas Works, into a canal. The confinement of river to a narrow bed would carry silt well down the harbour and prevent its lodgment about the shallow ports at the upper end."[43]

In 1878, British engineers Walter Kinipple and William Morris submitted a proposal for a stone dry dock at the mouth of the Waterford River. Their proposal was as much about made ground as about the dock itself; Kinipple and Morris promised the project would involve a "reclamation of foreshore" from the gasworks to Job's Bridge at the river's mouth, which they estimated would take 100,000 cubic yards of material and create almost ten acres of new land. The material could come from dredging, dumping, or excavation, but it would pay for itself:

> The reclamation of the land between the Gas Works and Job's Bridge we include as part of the Harbor Improvements; and not only would it form an excellent place for the deposit of the excavated material, but would, in a very short time, become valuable building land and certainly produce an income more than sufficient to cover any deficiency in Dock dues in consequence of any bad fishing seasons.

"The construction of such important Works in the Harbor of St. John's," they added, in concert with Robinson, Hatton, and Harvey, "will necessarily raise the value of all the wharf property in Water Street."[44]

George Robinson, though, had additional thoughts on the question of the dry dock. In fact, Robinson had already issued a report on this very subject, after having been sent to evaluate dry docks in a number of cities on the American seaboard in the fall of 1882. Robinson met with operating engineers and considered modes of construction, expenses of maintenance, rates charged for repair, and condition and function – notably any springs or leaks, the challenge of securing a timber dock to a rock foundation, and the amount of coal needed to pump the dock out.[45] He concluded that the Waterford River was not an ideal location. Not only was

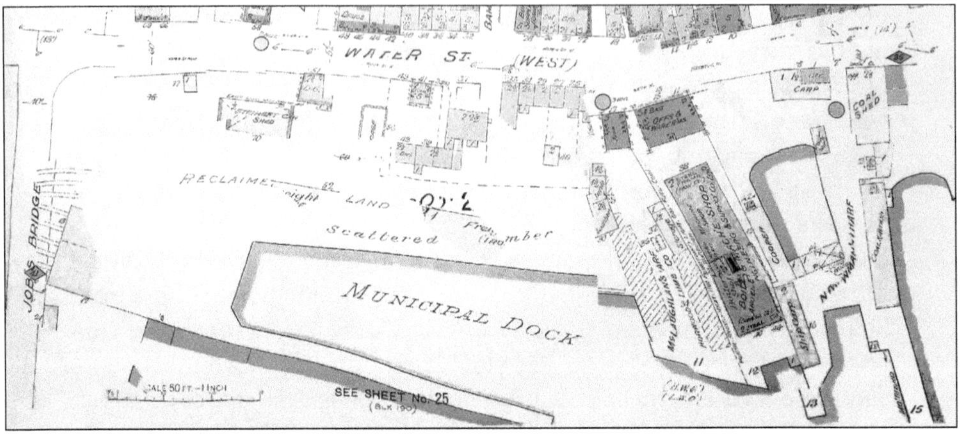

Goad Insurance Map, St. John's, 1893 (detail). The Charles E. Goad Company had supplied insurance maps in Canadian cities since 1875, akin to the Sanborn Insurance Company in the United States. Note the "reclaimed land" and the number of edits in a rapidly changing part of the waterfront. | City of St. John's Archives

he concerned about encroachment on "this valuable conservatory of tidal waters" (an interesting echo of the language at Saint John), he felt a dock should be located at the *other* end of the downtown, nearer the Narrows, which he felt was underused relative to the rest of the waterfront.[46]

But the weight of established infrastructure, fixed capital (to borrow David Harvey's phrasing), and political opinion all tilted toward the west end. Five months after Robinson's tour, the contract for a dry dock was awarded to J&E Simpson & Co. of New York, known for its timber dock construction. The contract authorized the company "to enter upon, take, use, and hold, any land, whether covered by water or not, and whether public or private, or any Beach, Shore, or side of the said Harbor (or any of the water of the said Harbor)," and to take stone, gravel, earth as necessary "over and through any lands intervening between the Dock and the Land."[47] The *Evening Telegram* pointed out that this meant the company could charge the government (and the public) both coming and going, for filling in the river mouth *and* for dredging material that had to be carted or boated away. The newspaper also raised concerns about how much valuable waterfront the dock might swallow.[48] By December 1884, though, it was finished, and St. John's was home to the largest dry dock in North America. Despite the *Telegram*'s skepticism, it was considered a feat of engineering to not only seal a graving dock of this size but to have remade the surrounding waterfront "filled with earth carted from the neighbouring hills."[49]

This was the paradoxical promise of Riverhead: it could extend the harbour *and* the waterfront at the same time. As Premier William Whiteway proclaimed once the dock project was underway, "The only direction in which [the harbour] can be extended is towards Riverhead. The opening up of this channel is therefore the first step towards extending the harbour." At the same time, filling the river flats would add new acres of land, "which in time must be valuable for business purposes."[50] Such statements, and the larger harbour debates of the 1880s, align with David Alexander's argument that this decade marked an important step in Newfoundland economic history, toward a policy of industrial resource development and infrastructure projects that would facilitate inland development.[51] Ironically, after all the concern about encroachment on the water, and all the interest in ship traffic and repair, the centrepiece of this "policy of progress" would be a railway across the island interior to open new terrestrial industries such as mining, agriculture, and forestry. Even the convoluted and combative political economy of nineteenth-century railway financing did not derail the shifting balance of power. The railway would become the primary occupant of Riverhead lands, with train yards, machine shops, and a new passenger station, and the Reid Newfoundland Company – that operated the railway – would be awarded the dry dock, as well, in 1898.[52]

The Foreshore Question

The new geography was, ironically, also a product of disaster. The bricks were still cooling from the Great Fire of 1892 – the largest in the city's history – when the legislature dealt again with the thorny issue of land tenure. The "foreshore question" had never been resolved but was now acute in the burnt-out areas along the waterfront where many leases were about to come due.[53]

The fires of 1846 and 1892 supplied bookends by which to measure the tenants' predicament: leases that did not compensate any tenant improvements had proven a strong financial disincentive for spending on better construction. Donald Morison, member for Bonavista, offered the House of Assembly an illustrative example:

> He knew of a case where the waterfont [sic] had been taken by the landlord. About the time of the fire of '46, a gentleman in town took a lease of a waterside premises in the neighborhood of Messrs. Ayre & Sons. At the time the water came up very near the back of the house, but the tenant

H.M.S. Blake, from Halifax.

THE FI

FROM A SKETCH B

At a meeting of the Mansion House Committee of the fund for the relief of the distress occasioned by the disastrous fire at St. John's, Newfoundland, the Lord Mayor presided, and there were present Sir Terence O'Brien (Governor of Newfoundland), Sir Charles Tupper (High Commissioner of Canada), the Hon. A. W. Harvey, of Newfoundland, and others. It was resolved to remit a second sum of £9000 to Newfoundland by telegraph, and to request Government to send a company of Royal Engineers to assist in laying out the rebuilding of the city, if acceptable to the Newfoundland Government.

Our View of the scene of havoc and desolation, when the great conflagration had almost stopped, is supplied by the sketch received from a correspondent, Mr. H. F. Bradshaw, whose letter is dated July 13. To the right hand, on the north shore of the harbour, and on the slope of the hill, upon the summit of which still stands intact the Roman Catholic Cathedral with its twin towers, as far as the small open space to the east, with the Methodist Church in Cochrane Street, also uninjured, rising above, just within the limits of this view, all is a mass of ruins, the chimneys of many houses remaining where the main buildings, usually constructed of wood, have been reduced to ashes. Smoke from the coal-yards on the shore, which continued to burn several days, partially obscures the prospect, different features of which at times became revealed to sight by the shifting of the wind. Below the hill, to the left hand, are the remains of several of the most conspicuous buildings—the fine English Cathedral, the Presbyterian Church, the Athenæum, and the Methodist Church in Gower Street, which have been destroyed. This was the central quarter of the city. The fire began at five o'clock in the afternoon of Friday,

PLAN SHOWING THE PART OF

THE GREAT FIRE IN THE CIT

"The Great Fire in the City of St. John's, Newfoundland," | *Illustrated London News,* 30 July 1892.

DOWN.

F. BRADSHAW.

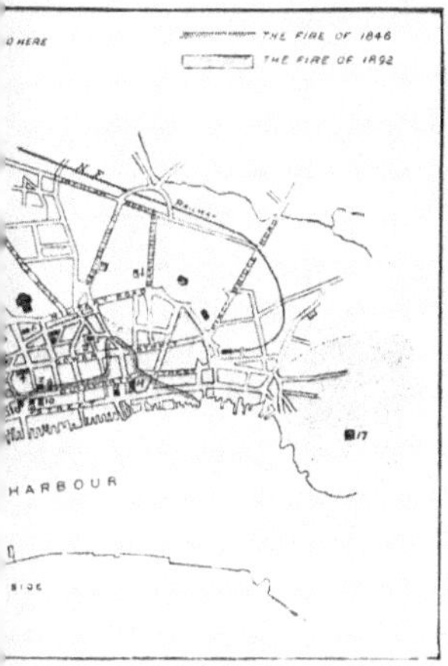

July 8, high on the western side of the hill, and was driven down, by a fierce wind from the west-north-west, sweeping over both the lower ground and the hill-slopes comprised in this description, sites densely covered with houses; while that portion of the town lying farther west, up the harbour, was spared, as well as the opposite shore. Near the southern shore, as seen to the left hand in our correspondent's sketch, lay on July 13 the ships which had then arrived from Halifax on a mission of relief, H.M.S. Blake and H.M.S. Emerald; also the screw-steamer Ulundi, with relief stores, is lying in the mid-channel a short distance above. It is estimated that ten or twelve thousand of the population are rendered houseless and destitute. Efforts to provide for their wants are most imperative just now, but special appeals are made also for aid in rebuilding the English Cathedral and the other churches, chapels, and schools, to which co-religionists in England and Scotland are invited to contribute. The English Cathedral was a noble Gothic edifice, designed by the late Sir Gilbert Scott, and built of stone, at intervals of time, during forty years past. The Church of England Synod Hall is also destroyed; the Methodist College, the Masonic Hall, St. Patrick's Hall, and the hall of the Orange Society. Meantime, the house-owners and proprietors of warehouses, shops, and other commercial establishments at St. John's complain that some of the owners of building land are disposed to exact a higher price for sites to be occupied by new buildings.

The hospital ship Albert, belonging to the Mission to Deep Sea Fishermen, has arrived at St. John's, Newfoundland, with a quantity of clothing for distribution among necessitous fishermen's families. The Governor's wife, Lady O'Brien, has kindly undertaken this task.

DESTROYED BY THE FIRE.

ST. JOHN'S, NEWFOUNDLAND.

> being in want of accommodation behind, proceeded to fill in, and on the ground thus made he built stores and other places of business. After the lapse of his term this made ground had to be handed over to the landlord which, he must say was an injustice to the city.[54]

While George Robinson had been primarily concerned about encroachment in terms of harbour function, those on the new city council and in the assembly tended to see the expanded waterfront in terms of property revenue. This position reflected the material reality that any lot of Water Street was now fully on and surrounded by solid, made ground that could be measured and modified, held by the commonwealth or taxed for its benefit.[55]

But Robinson's report from ten years before resurfaced in these debates. Edward Morris, the representative from St. John's West, decried the state of "public waters, hedged in as they are by a fence of public wharves." Morris, who wanted the assembly to establish a harbour trust to manage the foreshore, referenced Robinson's findings at length "to show the quantity of the public waters which have been filched from the public and converted into land by the throwing of debris into the harbour." For Morris, the absentee landlords were robbing the citizens of St. John's at least three times over: taking up public water, charging merchant tenants "exorbitant rents" on this new land, while "contributing not one cent to the city, the rightful owner of the property." And, Morris continued, "this process of encroachment is still going on, was going on to-day ... throwing debris into the waters to the harbour, making land for the absentee landlord at the expense of robbing the public." When challenged to support Robinson's decade-old calculations, Morris answered that he could find fifty witnesses to attest to the older waterline and the fact that the tide used to wash against Water Street.[56] No one could imagine that happening again.

Postscript: Past as Prologue

The old waterline may have survived in scattered memories and street names, but at the turn of the twentieth century St. John's did not consider itself bound by history. On the face of it, the city was encircled by the same topographical, climatic, and economic constraints. It counted only thirty thousand residents, its population having plateaued for decades.[57] Even today, the rocky walls of what nineteenth-century visitors called the "mighty Coliseum" are still intact and winter temperatures do last "to

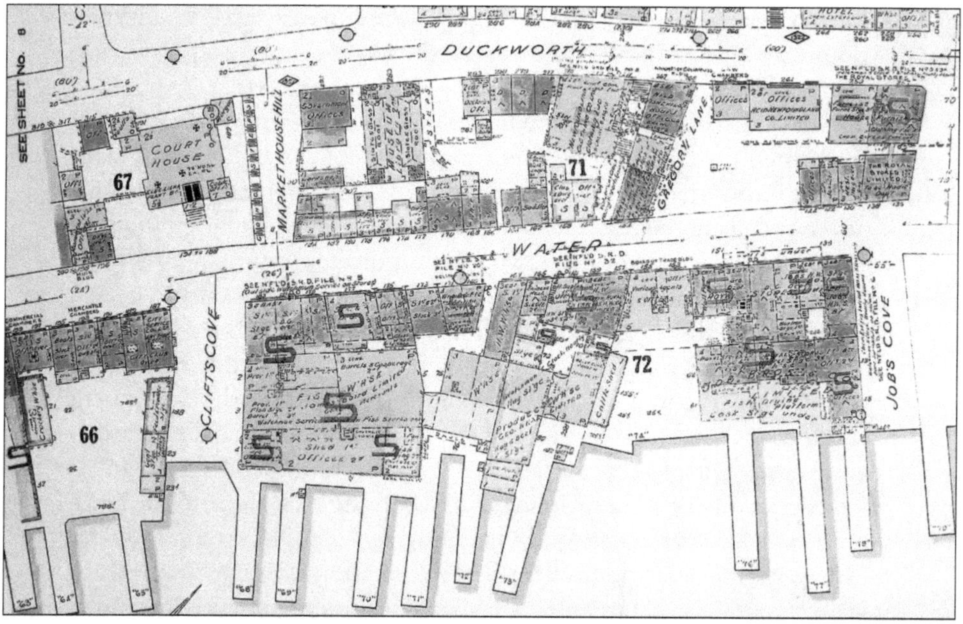

Underwriters' Survey Bureau Limited, Insurance Map, St. John's, 1946 (sheet 9). | City of St. John's Archives, courtesy of Verisk Canada

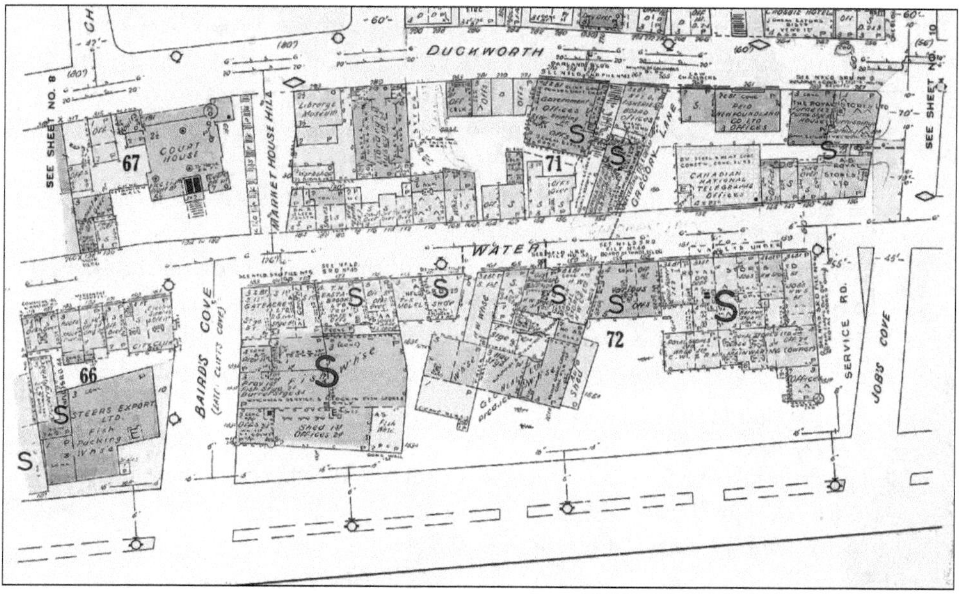

Canadian Underwriters' Association, Insurance Map, St. John's, 1967 (sheet 9). | City of St. John's Archives, courtesy of Verisk Canada

April! – or May!" (or June, for that matter). Resource extraction continues to dominate the provincial economy, and the offshore industries, especially, are still riven by disaster and the tumult of global markets.

But the encroachments of the long nineteenth century both signalled and affirmed the city's commitment to projects of extraction and expansion. As one booster wrote in 1899, "The far off and apparently impossible is now our daily heritage."[58] Indeed, the direction of the twentieth century would follow the tracks laid down in the nineteenth, on an ever more ambitious scale. The turn to road and truck transportation only increased demand for solid waterfront space. Finger piers were replaced with long squares of asphalt that quite literally cemented and extended the older made ground. The result is a "harbour apron," a solid swath of pavement that enclosed the linear piers and realigned harbour traffic parallel to rather than inserted with waterfront storage. The apron, like the TransCanada Highway roadwork at Saint John, also folded the harbour into the automobile era, with parking lots and an arterial Harbour Drive (built 1959–64) to deliver drivers into the city centre from the suburbs in hopes of making downtown shopping competitive in an era of suburban malls; the kind of access route that Halifax would avoid by cancelling its own proposed Harbour Drive. The waterline is now pushed further out and cordoned off by a chain fence, separating the docking berths of the Port Authority from the sidewalk of Harbour Drive. Harbour Drive rejoins Water Street high above the property of A. Harvey Co., a company that dates to the eighteenth-century Atlantic trade and now supplies heavy construction equipment and the apparatus (fuel, salt) for trucks in the same location as its nineteenth-century premises.[59] A thinner apron of asphalt and gravel runs the entire length of the south side and into the Narrows. This is a waterfront largely enclosed *from* the city.

Nineteenth-century observers would find the presence of bulk fuel on the south side more familiar, though cod and seal oil have given way to petroleum. Indeed, the most eye-catching feature of the St. John's Harbour today is not Cabot Tower on Signal Hill but the white Irving Oil tanks looming high above the city on the south side hills. With the discovery of the Hibernia oilfield in 1979, Newfoundland turned firmly toward offshore oil exploration, which required the same set of facilities – refueling, repair, storage – along the waterfront. Whether in supply or materiel, this is now a petroshore. Literary scholar Fiona Polack points out that the branding of Newfoundland as an "energy warehouse" suggests resource commodities stockpiled for sale to outside buyers. I would add that "warehouse" evokes

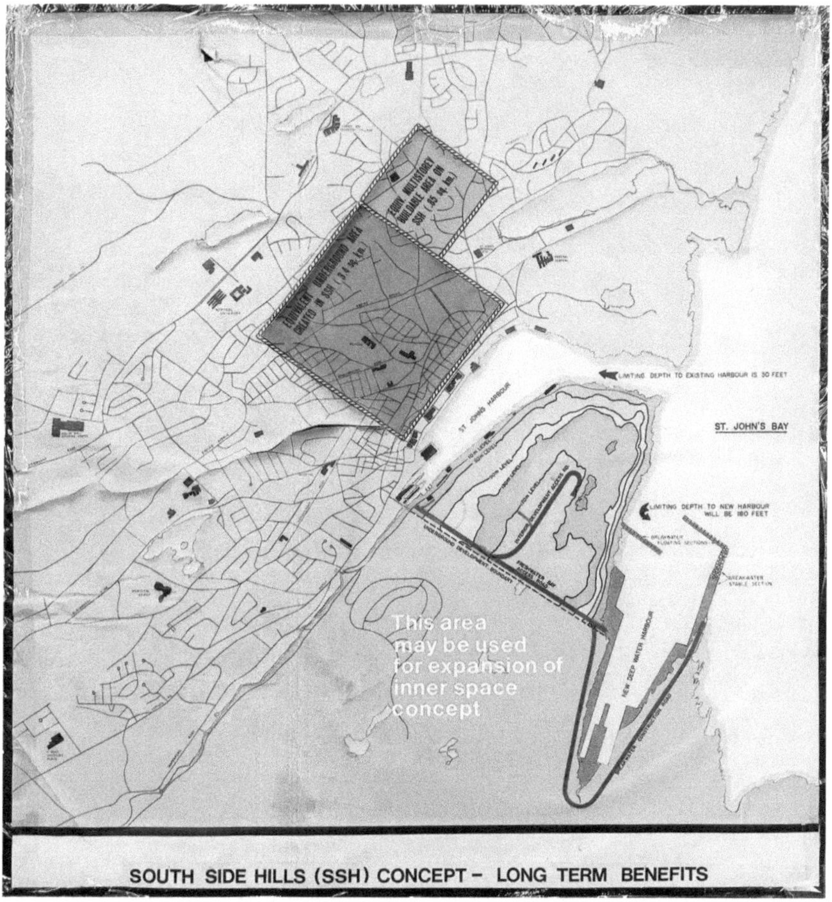

"South Side Hills" Concept Development (1970s). | Memorial University of Newfoundland, Queen Elizabeth II Library, Archives and Special Collections, Tom Kierans Papers, ASC-2014-0048

a very specific historical image, one that locates fuel economies at the waterfront but in appealingly nostalgic ways.[60]

The most important thread of continuity is conceptual: what historian Jerry Bannister called the politics of anticipation.[61] Modern Newfoundland has been defined by megaprojects, from Churchill Falls to Baie du Nord, that promise to break through the constraints of geography, the debits of history and hinterland standing, to empower and enrich the province. Development has grown larger, more remote, and infinitely more expensive but tethered by a faith in resource wealth and oceanic exploration.

View from Signal Hill (2023). | Photograph by the author

The spatial constraints of the harbour can be built beyond and through; in one dramatic example, the South Side Hills Development Corporation proposed hollowing out the far wall of the "mighty Coliseum" to create a massive new complex for storage, processing, and urban commerce – and a tunnel to a new industrial port at Freshwater Bay as large as the original harbour.[62] If William Whiteway believed the harbour could only expand inland, these projects have leapt imaginatively out of the harbour altogether.

The city's "daily heritage" is thus a paradoxical combination of this persistent belief in the "far-off and apparently impossible" and the different vulnerabilities in a new climate era created by past decisions. Measuring the history of encroachment is an important piece of adaptation to sea-level rise and storm surge. So, too, is observing the history of responses to disaster. Rebuilding after the fires (especially of 1846 and 1892) created an enormous financial burden for the city, but even that did not fundamentally reorient its economic profile or aspirations. There is an analogy to our present era here – that despite the mounting bills of climate response, we are unwilling to turn from the shorter profit horizon of resource extraction.

But along the St. John's waterfront there are glimpses of other kinds of possibility. In the spring of 2020, with the COVID-19 pandemic shuttering downtown commerce, the city turned Water Street into a pedestrian mall. (A departure from conventional thinking that better car access would improve downtown activity.) In a moment of acute anxiety, this invited people to return at least a little closer to the historic streetscape and waterfront at a human, pre-automobile scale. And if people cannot walk beside much of the harbour, they have always been able to climb above it. Signal Hill remains a cherished public commons and a core feature of the city's identity. It also allows us to see the harbour, albeit reduced, as whole and one. As Edward Morris said in 1892, "It is only by glancing at such an object lesson as this that we can appreciate the value of what we are now seeking to recover."[63]

3
Nature Opposed
Charlottetown, Prince Edward Island

Today Prince Edward Island, lying low in the Gulf of St. Lawrence, is one of the most dramatic harbingers of climate change in the North Atlantic. With each storm surge, more of the Island's signature red soil dissolves outward with the tide. Estuarine rivers threaten to fracture the island into a micro-archipelago, changing the Island's entire composition of land and water.

This is not our usual image of the Island, to be sure. Dubbed "the Garden of the Gulf" in the late nineteenth century, made famous by *Anne of Green Gables*, and celebrated in tourism campaigns, the tranquil landscape of Prince Edward Island signals to many of us the appeal of rural stability and security. But the proximity of coastal erosion on a small island now makes slogans like "you're never more than fifteen minutes from the beach" feel a bit more like a threat than a promise. Satellite images show how far water penetrates inland, making the Island's terrestrial integrity seem surprisingly vulnerable. Photographs of the north shore calving into the Gulf of St. Lawrence and rivers flooding red into the Northumberland Strait have begun to unsettle the portrait of rural permanence. It feels as though we are witnessing an acceleration of historical time, with the contours of the Island changing even as we watch. Living with the coast feels both urgent and uncertain.

At the same time, the relentlessly rural gallery of Island images – lupins and furrowed fields and piles of lobster traps – belies the fact that the Island, like the rest of North America, shared in the urban turn of the industrial era.[1] To be fair, Charlottetown is Canada's smallest provincial

PEI from the International Space Station (2015). | NASA Earth Observatory

capital, and (at about only forty thousand residents) does not even rank in the top one hundred largest cities in the country, in a province with the highest proportion of rural residents. Yet a town that registered a few hundred inhabitants at the start of the nineteenth century saw accelerated growth over the course of the nineteenth and early twentieth centuries, in real numbers and as a percentage of the Island's overall population.[2] Even in the era of Avonlea, with its portrait of pastoral prosperity in sharp contrast from most of the Atlantic region, the Island was never cut off from the wider trends of the North Atlantic world, including urbanization, rural outmigration, and industrialization. As Edward MacDonald has observed, "To call Prince Edward Island a pre-industrial society during the first half of the 20th century would be to ignore the degree to which the machinery of Island life had engaged with the urban-industrial complex that gave shape to the century."[3]

Enacting that urban-industrial complex in fact began decades earlier in the mid-nineteenth century, as evinced by Charlottetown's relationship with its water. In what follows, I trace how a colonial city initially praised for and defined by its aqueous features gradually erased the two wetlands framing its downtown core. Spring Park Brook supplied a corridor of fresh water that, divided among many users, remained something of a civic asset even as it was diminished; the East Marsh was an ill-defined (indeed, generally unnamed) wetland hardened and absorbed into industrial fabric and largely excised from civic life. Both complicated urban ambitions by creating spaces that did not always align with expectations of urban land and waterfront, and both accordingly became the subject of significant civic debate and intervention. The decisions regarding these two watercourses over the long nineteenth century profoundly influenced the shape of the city and its sense of possibility.

In this, Charlottetown is not particularly unique; covering fresh water and filling in salt marsh were fundamental to the entwined arcs of landmaking and city-building in nineteenth-century North America. But the story of Charlottetown asks us to broaden our conception of the urbanized coastline beyond the harbours of London and New York.[4] If even the small capital of a largely rural island – far from our image of metropolis, or industrial actor, or even a city – bears marks of the same commitment to development and growth, we may begin to appreciate how pervasive these forces have been in shaping the landscapes we inhabit. Identifying the tendrils of industrial infrastructure in even a small city (no more than nine thousand residents at the time of Confederation) also unsettles the bucolic Island of popular memory and the Atlantic region's resource-rural image within the federation. Despite its small size and the framing Island pastoral, Charlottetown was, like the other cities in this book, designed and developed in an era shaped by ambitions of growth, capital, and global connection. It is a poignant reminder of how much water runs through, and beneath, our coastal cities, as well as alongside them. And finally, it carries a lesson about the literal and metaphorical weight of past decisions – and the ways our choices today are framed and, more to the point, limited by past interventions.

Colonial Charlottetown: A City of Sea Creeks

It is important to acknowledge that the colonial county seat as imagined and planted in the 1760s sat within older Mi'kmaq and Acadian geographies that had long made use of littoral or coastal space. The major rivers crossing the island already known as Epekwitk and Île Saint-Jean were essential transportation corridors, while the coastal wetlands and estuaries

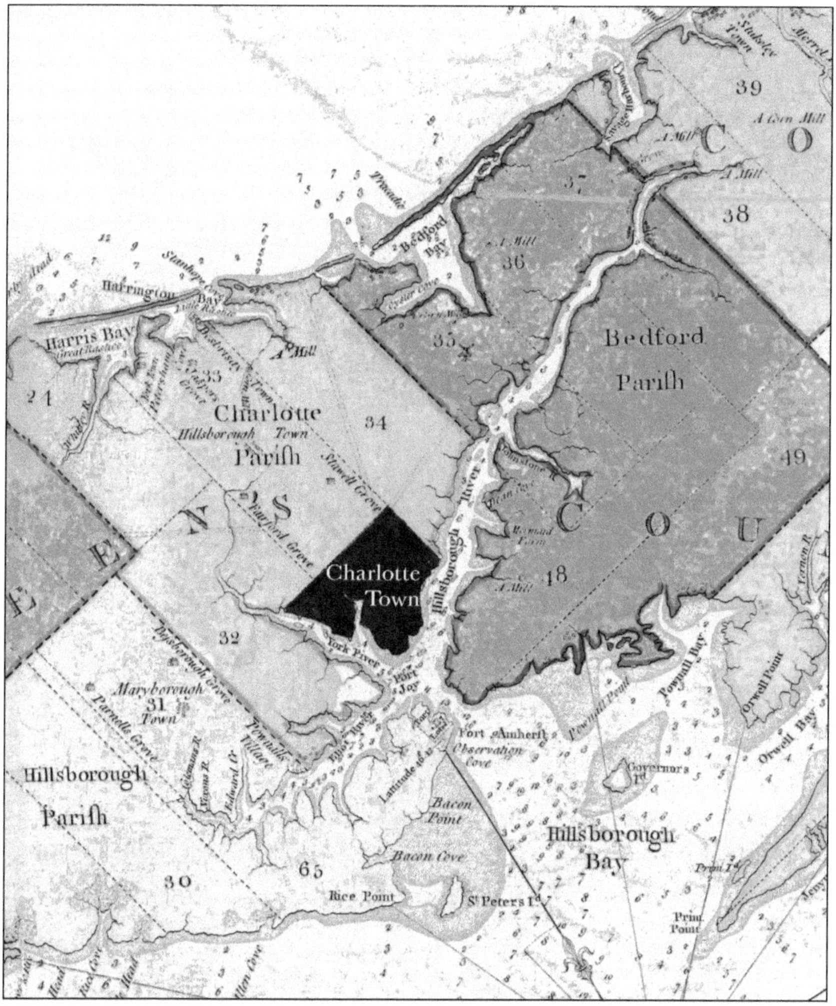

Samuel Holland, *A plan of the island of St. John with the divisions of the counties, parishes, & the lots as granted by government, likewise the soundings round the coast and harbours, published in 1775* (detail). | Norman B. Leventhal Map and Education Center at the Boston Public Library

were prized for fish and shellfish harvesting and for farming the tidal marsh and low-lying riverbanks. As Samuel Holland noted, "The Rivers are properly Sea Creeks, the Tides flowing up to the Heads, where Generaly [sic] Streams of Fresh Water Empties themselves."[5]

Holland was tasked with surveying the island from stem to stern in the wake of the Seven Years War, and his work – culminating in *A plan of the island of St. John with the divisions of the counties, parishes, & the lots as granted by government, likewise the soundings round the coast and harbours* – is one of the best examples of cartographic repossession from the imperial era in the region. Holland divided the island into three counties for governance and sixty-seven lots for property, as well as three plats for county seats. Charlottetown, named for the reigning queen of the United Kingdom, was to be placed in the middle of the Island as the seat for Queen's County and would be the only one of the three seats that observers considered to *be* a town for some time.[6] Holland's plan located Charlottetown on a small south-facing peninsula across the harbour from the Acadian settlement of Port-La-Joye, razed and rebuilt as Fort Amherst.

What is striking about Holland's description of the chosen townsite is how much of his evaluation centres on water, of different kinds, for different purposes. For one, the harbour is not a singular, enclosed body but really an intersection of waterways, leading out to the Atlantic and inland through the Island:

> The Capital called Charlotte Town is proposed to be Built Upon a Point of the Harbour, Betwixt Yorke and Hillsborough Rivers, as being one of the best and nearly a Centrical Part of the Island, has the advantages of an Immediate and Easy Communication with the interior Parts of the Island, by means of Three Fine Rivers, Hillsborough, Yorke and Elliot.

Such a location promised the most secure and strategic command of the harbour itself and any movement inland. On the townsite itself, there was also, crucially, a source of fresh water:

> The Ground Designed for the Town and Fortifications is well situated upon a Regular Asscent [sic] from the Water side; a fine Rivulet will run thro' the Town, a Battery or Two some Distance Advanced, will Intirely [sic] Command The Harbour.

Holland emphasized that the numerous waterways here served as security for any settlement:

> An Enemy attempting to Attack the Town, cannot do it without Great difficulties, Viz. having Passed the Batteries at the Entrance of the Harbour, they must attempt a Passage up Hillsborough and York Rivers, the Channels of both which are Intricate, and the Entrances of the Respective Channels will be so near the Town that it must also be Attended with the Greatest Hazard.

Finally, the expansive harbour would shelter a commercial fleet, supplying the town's economic purpose, since the established fishery remained anchored on the north shore of the Island:

> As this Side of the Island cannot have any Fishery it may probably be thought expedient to Indulge it with some Particular Privileges, and as all Judicial and Civil as well as a Good part of the Commercial Business will be Transacted here it will make it at Least Equally flourishing with the other County Towns.[7]

Between the "Rivulet," the three rivers, and the harbour, Charlottetown seemed to promise water at all scales and for all needs.

The first plan for the town itself was issued in 1768 by Charles Morris, surveyor general of Nova Scotia (the headquarters of British military presence) and refined by Thomas Wright (who would become surveyor general of the Island proper) in 1771. The plan followed the standard ordered arrangement of the "model town" being planted up and down the British-American seaboard.[8] Straight and spacious streets form regular blocks and property lots, with central space reserved for the institutions of governance and faith, together with a commons and a "royalty" beyond the grid for pasture. But as Holland's initial description suggests, water frames Charlottetown's plan, in both intentional and unexpected ways. Even Morris, who was blithely assigning identical plans throughout Nova Scotia and New Brunswick, here conceded the brook to the west and the salt marsh to the east and squeezed his town plat in between. (Morris also noted "Boggy Ground" at various points around the would-be city blocks.) Wright labelled these as West Creek and Charlottetown Creek, suggesting discrete and defined waterways, but neither name really stuck. West Creek was renamed Spring Park Brook, and Charlottetown Creek was lost in the marsh.

The environmental realities of the nineteenth-century town soon belied the antiseptic ideal of the eighteenth-century plans. "The town has kept pace with the country," one observer wrote in 1851, meaning that Charlottetown and its hinterland were meeting imperial expectations for both settlement and export production. With the coal seams of northern Nova Scotia and Cape Breton only a few hours away, steamers were plying the Northumberland Strait from Charlottetown to the port of Pictou by the 1840s, while wharves were extended further and further out across the tidal flats to accommodate the Island's export trade and "bring its fruits to the world's market."[9] Charlottetown's incorporation as a city in 1855 was

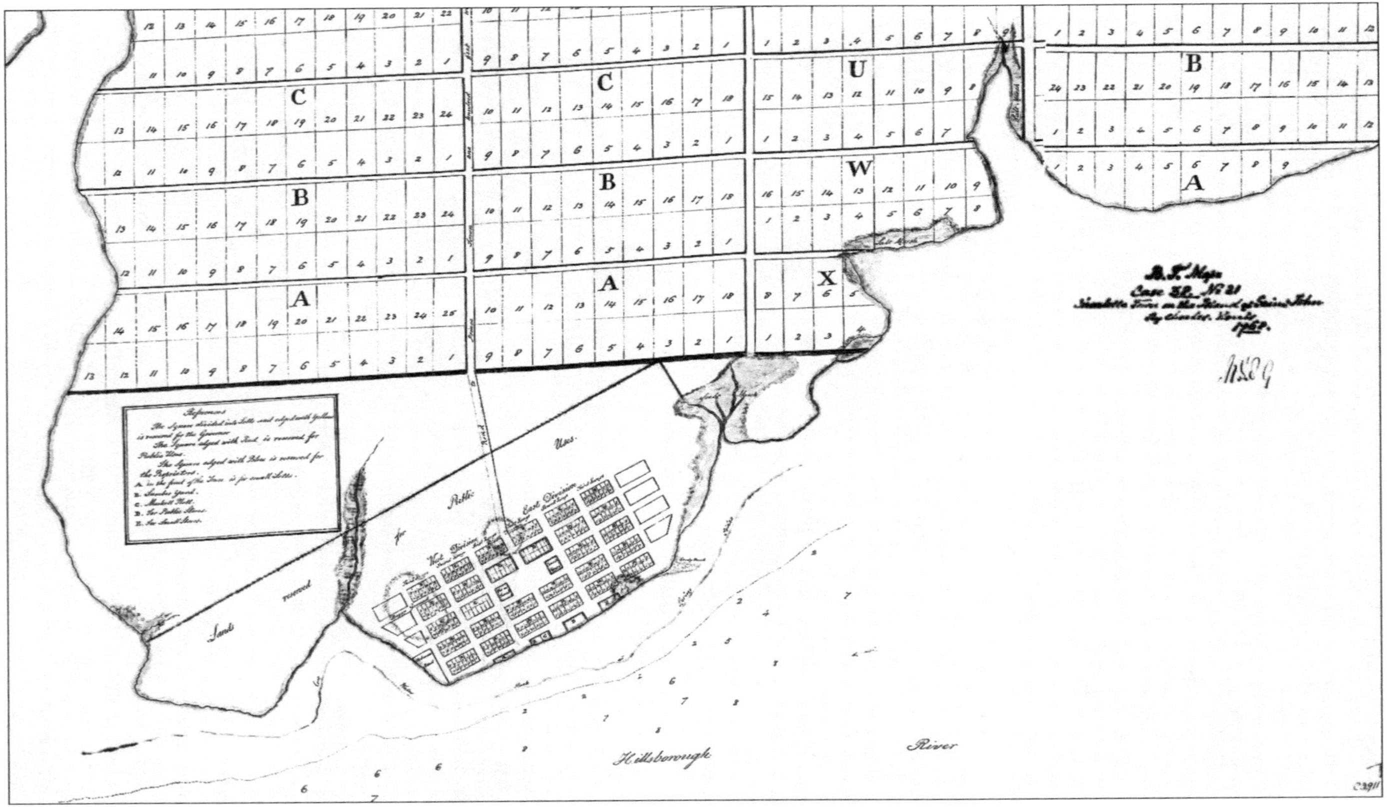

"A plan of Charlotte Town on the Island of St. John, and the pasture lotts and reservations etc., by the order of the Honourable Michael Francklin Esquire, Lieutenant Governor of the Province of Nova Scotia. Done by Charles Morris, Chief Surveyor, June 1768" (detail). | Library and Archives Canada, H1/240/Charlottetown/1768 an administrative acknowledgement of the demands of a growing centre, especially demands on water. As in other nineteenth-century cities, fires were a sporadic reminder of the need for a ready freshwater supply; pollution, health, and ambiance a daily one. (Duncan Campbell, writing a history of the Island in 1875, admitted, "The redolence of Charlottetown can hardly with truth be said to be elysian."[10]) The water that Samuel Holland had found so ubiquitous, industrializing Charlottetown would find increasingly inconvenient.

Spring Park Brook and the West Bog

For these reasons, the creek that formed the original western boundary to the town in the later eighteenth century became central to its politics by the middle of the nineteenth. Now renamed Spring Park Brook – a more bucolic name taken from the suburban estate it ran through north of town, and one considered more fitting for the "villa" lots being divided along its course – it originated in the old Royalty and ran southwest for about four kilometres to the harbour, draining most of the town's north and west in its watershed. The brook, and especially its source, still safely removed from most urban congestion, was briefly touted as a possible source of clean water for Charlottetown residents, who otherwise relied on wells and pumps throughout the town, often polluted by household waste.[11] But nimble industrial entrepreneurs moved in before the nascent civic bureaucracy could carry water away.

Unfortunately, if typical for nineteenth-century cities, this industry included particularly noxious establishments: tanneries, lime kilns, a distillery, and a woolen factory that was charged with rendering the stream unfit for man and beast. The low topography did not help. As the Charlottetown *Examiner* explained in 1886 (to readers who were likely all too aware of the situation), "The drainage from a large area of low, swampy land flows, or is conducted by drains, into the Spring Park stream, which itself is a continuation of a muddy stream descending from a muddy stream far back in the Royalty."[12] Unsurprisingly, the next year the city decided to locate its first waterworks further back in the Royalty, on another stream.[13]

As the brook crossed Green Street and then Brighton Road, less than three hundred metres from the harbour, its function changed. It had long served as an unofficial boundary between the town plat and green space reserved for the governor's residence and farm. Water, of course, has served as a border since before Moses parted the Red Sea, and industrializing

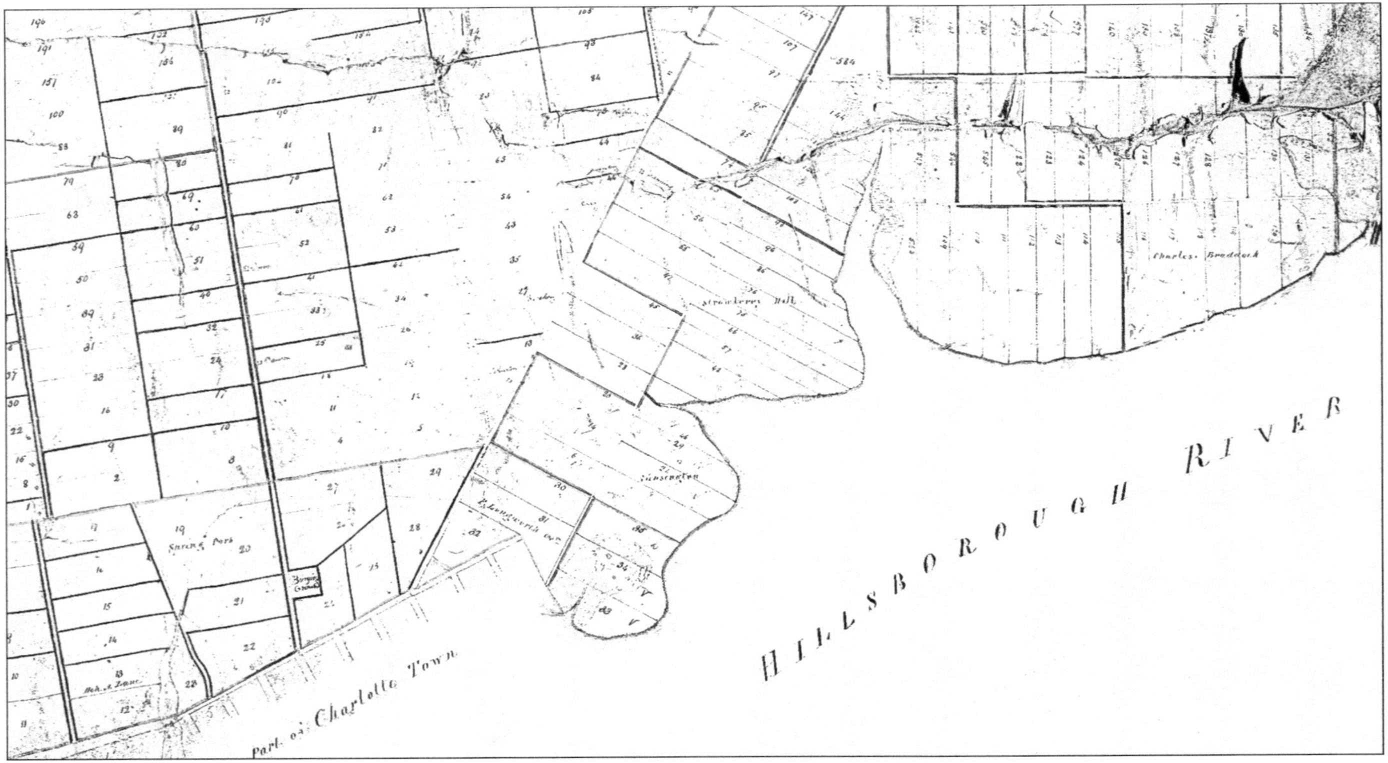

George Wright, "Plan of the Royalty of Charlottetown from a survey taken in the year 1831 by Order of His Excellency Lieutenant Governor Ready" (detail). The Royalty assigned lots for pasturage and agriculture outside of the town proper. Both Spring Park Brook (left) and the "Kensington Bog" (right) are visible in this section of the map. | Island Imagined Map Collection/Robertson Library, UPEI and Public Archives and Records Office, Prince Edward Island, no. 1163 cities frequently took advantage of the depressions and boggy inconvenience of runs and wetlands to spatially demarcate segments of urban life: public and private, residential and industrial, desirable and less so. Accordingly, as Spring Park approached town and harbour, public debate focused less on the quality of water for supply and use and more on the presence of water and the character of the urban landscape surrounding it.

At the heart of this concern was the lower stretch of brook, where it widened and then entered the harbour. Known here as Governor's Creek or Governor's Pond for its location alongside the governor's property, the water was crossed at Brighton Road by "Black Sam's Bridge" named for Samuel Martin, a free Black labourer, and at the harbour end by a bridge guarded by sentry. In this, the two ends of the pond represented the socioeconomic poles of Charlottetown. By the time of the city's incorporation, some were clamouring to make Governor's Bridge part of a "public walk or promenade" or a route through a new city park next to the governor's estate.[14] At the same time, the east shore of the pond, now tagged "the Bog," housed a working-class community of Black and Irish Islanders.[15]

The short distance between these two poles – neither was out of sight of the other – helps explain why public opinion about the space was so intense. By the 1880s, as water quality downstream of the tanneries and the like diminished, Charlottetown newspapers devoted more and more space to the question of what to do about the pond. Those who hoped to see a park, promenade, or posh housing more inviting to the better sort derided the pond's eastern edge as a "cesspool," "eyesore and disgrace to the city," "that filthiest of holes." Without appropriate intervention, the newspapers warned, "fever, mud and immorality will still confound the inhabitants of the West Bog." Notably, there was more public condemnation here than for the tanneries and other industries, which were tolerated for their economic contribution to the town.[16] In the emerging modern era, marshes and wetlands – despite their remarkable ecological productivity – were generally seen as an impediment to cultivation or construction, a source of unhealthy fumes, a challenge to the project of progress. When white Islanders conflated a racialized community with an undesirable landscape ("bog" is rarely a complimentary term), it only amplified the marginalization of both. Pollution meant the visual pollution, the "continual offence to the eye" of "shanties" that, it was feared, reflected more poorly on the city's moral standing in the eyes of visitors than waste or bacteria in the stream itself.[17]

But critics could not always agree on the best course of action. Should "the swamp on the Brighton side" of town be drained, corralled, or

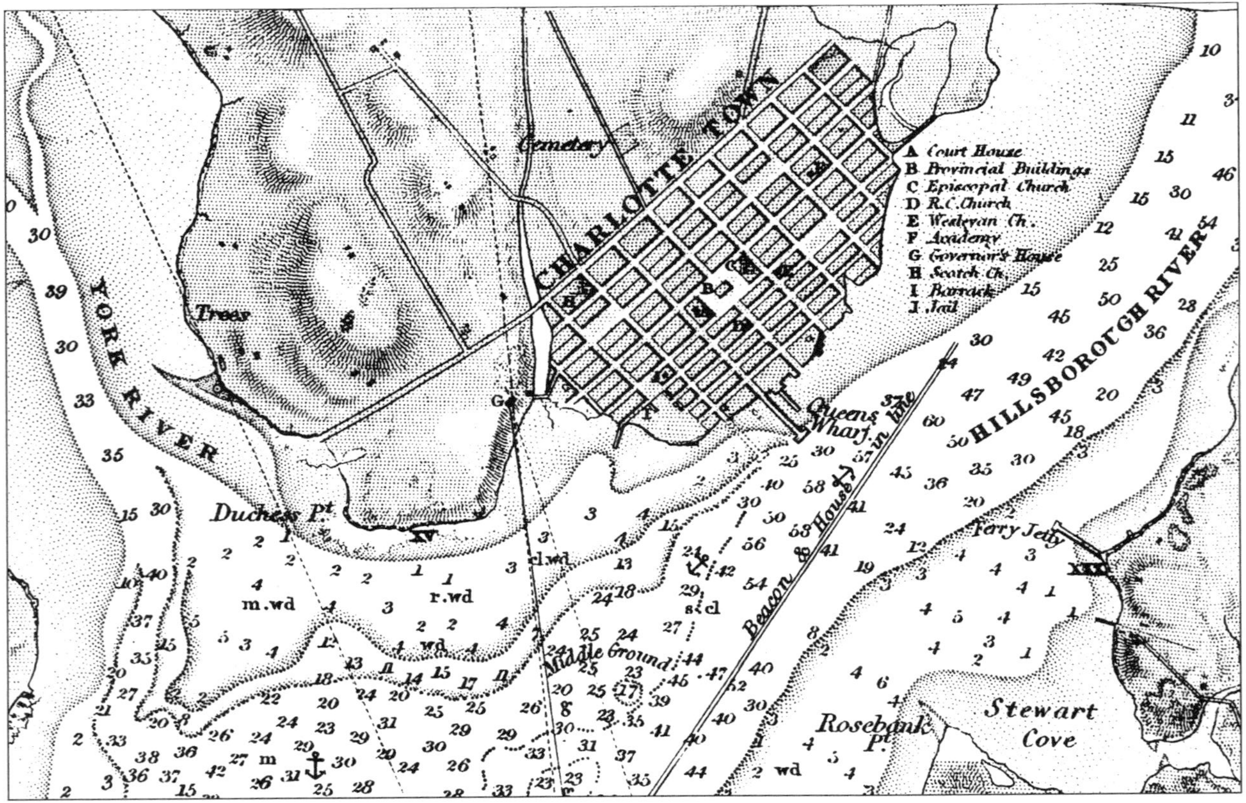

Prince Edward Island. Hillsborough Bay surveyed by Captn. H.W. Bayfield R.N. (London, 1846) (detail). | McGill Library, Rare Books and Special Collections, McGillLibrary-rbsc_map_bayfield-071-19174, archive.org flooded?[18] Should there be a public esplanade or waterfront properties for the leisured classes? What they *did* agree on was that a riverside redeemed would improve both the city's reputation and its property values. In 1876 one writer suggested that "by taking down some of the small houses in the neighborhood, and altering Government Pond into a beautiful stream, running through the bridged gardens and lawns behind residences situated on the present site of the West Bog – this portion of the town will be the most beautiful part of it."[19] Ten years later, architect W.C. Harris offered an elaborate plan that occupied most of the *Examiner*'s front page. Harris was just as invested in the gentrification of the area as most of those who appeared in print, but he presented his case through hydrological logic. He pointed out that trying to drain the pond would be futile; it would be refilled either by water at high tide or surface water flowing into the depression of the "swamp." "Nature," he said, "thus seems to be very strongly opposed to the idea of draining the pond."[20] A much better idea was to contain and beautify the water *as* a pond for the "improvement" of the neighbourhood.

As Harris's observation suggests, urban water invites a battle of wills. Like many urban waterways, most of Spring Park Brook was lost not through a single dramatic shovelful of earth but a thousand small ones. North of Brighton Road, the brook was gradually covered to hide pollution and to facilitate the expansion of roads and suburban neighbourhoods.[21] In contrast, the pond at the brook's mouth *was* preserved, although not really out of nature's own opposition and certainly not through any human intention. Its harbour edge was thickened with a causeway and then the long-desired esplanade by the end of the century, supplying a gateway to Victoria Park, which had been established in 1873 and which remains the city's primary greenspace. The waterfront was arranged as an upscale subdivision and the Bog neighbourhood gentrified and absorbed into residential blocks. The pond survived until the 1960s, when it was almost entirely filled in as a parking lot. But, nevertheless, water persists. A cluster of cattails at the edge of the parking lot marks the last of the pond, and the brook occasionally remerges during the spring freshets to flood the namesake Spring Park Road.[22]

The East Marsh

Ironically, the watery place that Thomas Wright had named "Charlottetown Creek" was never really integrated into the city of Charlottetown. If Spring

Aerial Photographs, 1958 (left) and 1990 (right). | Reproduced from the collection of the National Air Photo Library with permission of Natural Resources Canada and the Province of Prince Edward Island

Park Brook invited many uses, the marshland on the eastern edge of the town plat seemed too ill-defined to be directed into civic service. Without clear distinctions between water and land, or between the tendrils of fresh water draining toward the harbour and the saltwater intrusion in the low land, it wasn't clear what could be done here. But when the marshy shorescape that refused to cooperate with the town grid was eventually filled, expanded, and remade, it could finally be put to use. This meant dispelling the changeability of marsh for the solidity of a rail line, supplanting the energy of tidal shore with that of fossil fuels, and confining wetland to the point of erasure. The transformation of the eastern marsh irrevocably altered the city's shoreline in ways that reveal most clearly not only the material reach but the ideological momentum of the industrial age and the particular demands of railroad infrastructure.[23]

For at least a half-century the marsh remained intransigent. In 1842, naval surveyor Henry Bayfield still found it a force to be reckoned with, halting the expansion of city streets. The town lots bordering the marsh remained largely empty – a sure cartographic cue of a changeable, difficult, and probably soggy surface. But pragmatic Charlottetown had begun to use this wet periphery as a border space, to distance the sprawling or unappealing elements of the urban fringe from its residential neighbourhoods. The coal-powered gasworks of the Charlottetown Gas Light Company was built on the edge of the marsh in 1854. A large government-run stock farm pastured cattle on the salt marsh hay – cattle destined for an abattoir also on the edge of town.[24] The Royalty edge offered more semi-rural scenery for suburban villas and buffering distance for the colonial fairgrounds, a shooting range, and an insane asylum. By the later nineteenth century, the government began to transfer Crown shoreline to private hands.[25] But unlike Spring Park Brook, which held enough public value to draw public attention, the eastern marsh seemed destined to be left to industrial rather than civic purpose.

After 1871, fossil fuels remade the full length of the Island with the building of the Prince Edward Island Railway. The line took a controversial, distinctly nonlinear and, accordingly, rather costly route across the Island. As the colony's fiscal reach exceeded its grasp, it faced the equally unappealing prospects of bankruptcy or Confederation with Canada. We know how *that* story went: Prince Edward Island joined Confederation in 1873, the federal government assumed responsibility for the debts and for operating the Island line along with the Intercolonial Railway, and in 1918 the Island Railway was folded into the Canadian National Railway's network. Though we think of them as linear infrastructure ("laying down tracks"),

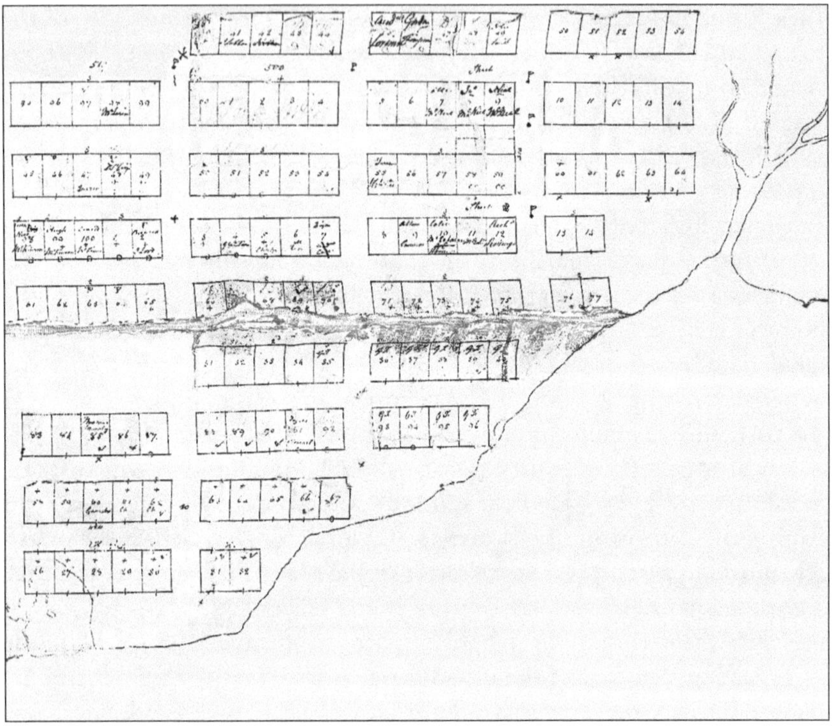

"The Original Plan of Charlottetown" (ca. 1780) (detail). Note both the dotted promontory of shallows at the bottom of the map and the multifingered watercourse – a sign of a generally wet area – at the top right. | Island Imagined Map Collection/Robertson Library, UPEI and Public Archives and Records Office, Prince Edward Island no. 4233

railways require an enormous amount of earth-moving and earth-enclosure for fuel, storage, and maintenance. On a small island riven with "sea creeks," the railway would also have to contend with water.

On PEI, the railway ran a spur down the eastern edge of Charlottetown before curving into the downtown, to the railway station at the foot of Weymouth Street and the railway wharf. Building the trestle out into water had the effect of demarcating different segments of the tidal shore: there was "the pond," now more clearly carved from the Hillsborough River; the water that insisted on seeping into city blocks and buildings; and the marsh beyond. Residents and visitors alike complained about the "water-soaked and slimy" site of the station, as a(nother) threat to public health, decency, and the city's good standing. The family that ran the Railway Hotel on Water Street reported their cellar held several feet of stagnant water, thanks to "low land adjoining the house, which land was receptable for sewage from the neighbouring premises."[26] (Interestingly, Charles Morris and Thomas Wright had noted water at this same spot in their plans a century earlier. Sometimes it helps to look at old maps.) A few blocks east, things became even wetter, thanks to "the existence of a copious and rapid stream in the eastern part of the city, near the gas works."[27]

The newspapers presented a variety of "improvements" to address such "nuisances," including regular health inspections, breastworks across the mouth of the marsh (build one "and the marsh will dry up in six months," the writer promised), or a causeway that, the *Examiner* suggested, would facilitate a carriageway to the new driving park and exhibition grounds and at the same time "the salt water will be prevented from flowing in upon the marsh, thus ensuring its reclamation."[28] There was widespread acknowledgement that this was a watery part of the city, but unlike at Spring Park, there appeared to be little value in keeping it so. After Confederation, and the elevation of a colonial railway to a federal responsibility, there were semi-regular attempts to fill in the depressions at the railway yards.[29] But the construction of railway infrastructure would require shoreline redrafting on an entirely different scale and would permanently change the contours of the Charlottetown peninsula. It also tilted the waterfront away from seagoing traffic, drawing it into a terrestrial network more than a coastal one.

All of these became apparent in the debate over how, and where, to build a bridge across the Hillsborough River. The bridge was to extend the railway to Murray Harbour, on the southeastern edge of the Island. The city of Charlottetown wanted the railway to first turn west and connect to the other shipping wharves in the downtown core and for the bridge to begin from Prince or Cumberland Streets, or at least the railway wharf at Hillsborough Street. The federal government and its engineers wanted the bridge to cross further east of the city, at Kensington Range (a spot known as Shipyard Point), imagining a future expansion of the railway yards.[30] Both options assumed plenty of filling in and dredging out of the waterline but in different directions. The eastern marsh – which until now had been largely ignored as the city's marshy backyard – now fell squarely between the two camps.

It was not lost on anyone that this was a test of just how much weight Islander opinion carried in federal politics and, by extension, the value of a small eastern port in an increasingly westward-facing nation-state. The borders of watery places are always difficult to define, especially tidal ones; the lines of political jurisdiction (or credibility) in the wake of Confederation were similarly unsettled. City leaders emphasized just how watery

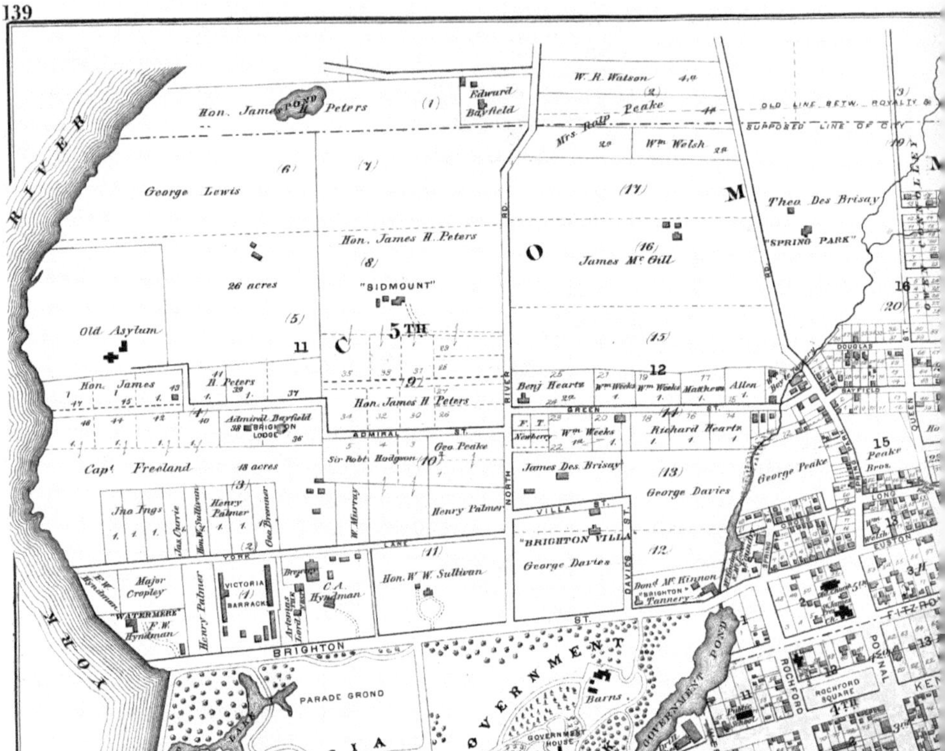

J.H. Meacham & Co., "Map of the City of Charlottetown," *Illustrated Historical Atlas of the Province of Prince Edward Island* (1880). | Island Imagined Map Collection/Robertson Library, UPEI

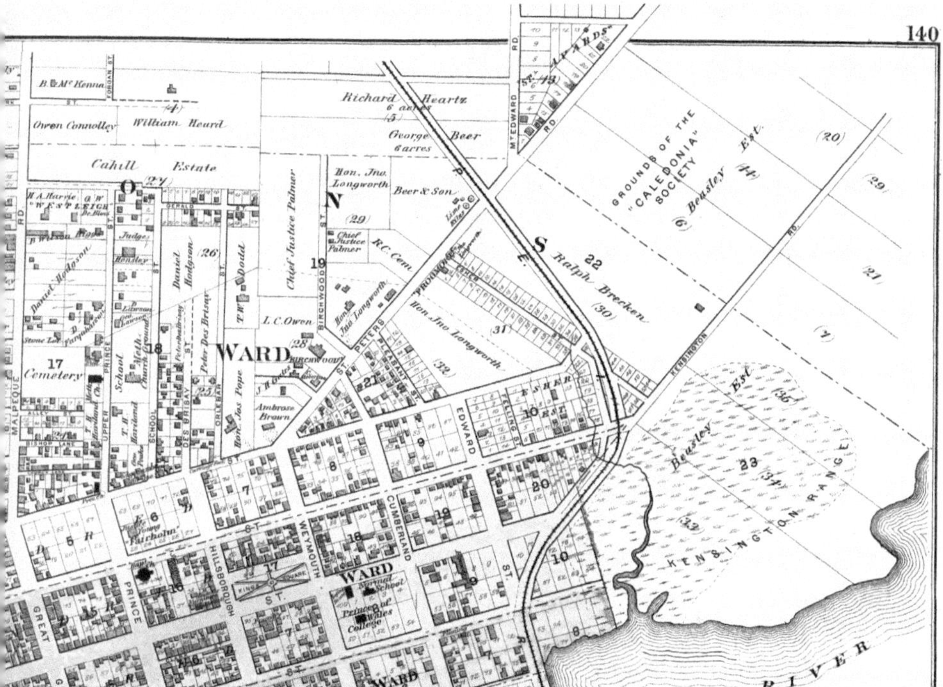
MAP OF THE CITY
of
CHARLOTTETOWN
P. E. I.
from actual surveys
Scale 400 ft. to one inch
C. R. Allen Surveyor
Note. Numbers marked thus 26 indicate
N^os of Common Lots
Numbers marked thus 20 indicate
N^os of Blocks used in assessment rolls
HILLSBOROUGH
RIVER
KENSINGTON RANGE
GROUNDS OF THE "CALEDONIA" SOCIETY
Cemetery
Gas Works
Station

and impractical the federal choice was, and newspapers lambasted the "stupid decision" to "locate the station on a bog," or "half bog half pond, and one member of the Board of the Trade states that he has repeatedly skated over it."[31] Even after Ottawa won out with the decision to build eastward, the question of what, exactly, was being built on remained contentious – and a way of asserting local knowledge against federal presumption. Edward Hackett, the Member of Parliament for West Prince, accused the Minister of Railways and Canals of doing little more than buying votes by hiring men to "fill in the hole" before an election:

> We have had winters for the past thirty years and we have had ice, and certainly if this hole required to be filled up it could have been done any time within the past thirty years. Now this little place at Charlottetown is well known to everybody ... That little corner between the railway and the gas house at Charlottetown has been overflowed with water for the last thirty years.[32]

Later the minister was asked to explain a fifteen hundred dollar expense "for land [the federal government] purchased from the province of Prince Edward Island ... near the end of the Hillsboro Bridge." The Member of Parliament for Queen's (who evidently could read a map), commented, "I do not think there is any government land there," to which the minister responded glibly, "It is land covered with water, I believe."[33]

If it *was* to be land, it had to be made so. By arcing out into the harbour, the railway had supplied a physical edge to a part of the water, which meant it could be understood as a discrete space. From there, it was easier to imagine that space as enclosed and infilled. The PEI Railway purchased the marsh pond and promptly invited the public to deposit their garbage in it – which, in fact, Charlottetown residents had been doing for twenty years, entirely in keeping with how industrializing cities tended to treat their rivers and wetlands.[34] By the summer of 1905, the newspapers reported that not only had Government Pond on Spring Park Brook been "cleaned up" but there was good progress being made filling in the pond by the railway station. (That said, there were still reports of drownings and near-drownings here for years afterwards, suggesting the water was not easily erased.)[35] After a fire that fall destroyed several of the railway shops, the federal government took advantage of the moment by filling in the rest of the pond and embarking on construction of a full complex of car shops and engine houses, sheds and wharf space, and a new station. To connect this new system to the Hillsborough bridge, fill first walled off the rest of

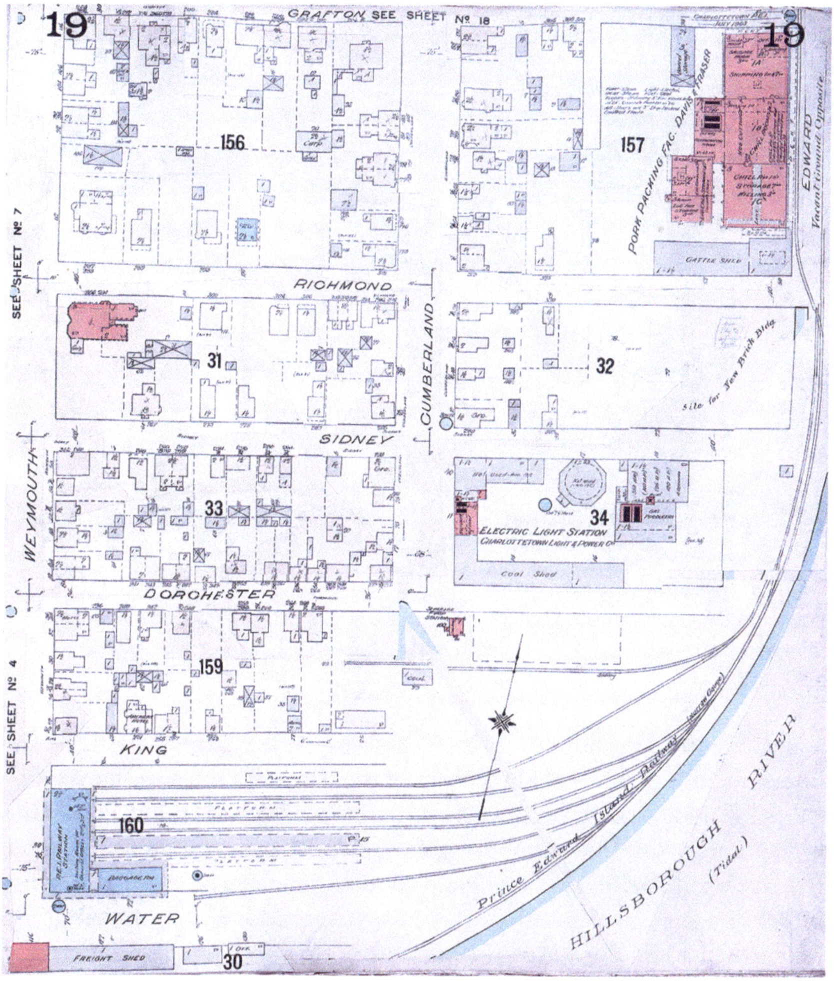

Charles E. Goad, Insurance Plan of Charlottetown (1903, revised 1917). The edges of blue that mark the original water line peek out from under correction slips pasted over the older plans. | Island Imagined Map Collection/Robertson Library, UPEI and Public Archives and Records Office, Prince Edward Island no. 4458_s6

the marsh from the harbour and then rail lines and roads leading out of the city were built. Such a dramatic and durable commitment to a single occupant led to the waterfront as the "geographically ordered, complex, composite commodity" envisaged by David Harvey.[36]

"Fancy what a different scene will be presented to the view in that section of Charlottetown when all these buildings are completed," declared the *Guardian*.[37] But "that section of Charlottetown" was now different not

Postcard of Charlottetown (ca. 1990?). The oil tanks are located on what will become Confederation Landing. | Prince Edward Island Museum and Heritage Foundation, XHF.2022.15.24

really because of the buildings – Islanders were entirely familiar with railway lines and light industry in this corner of the city – but the new shape of the shoreline. Not at all coincidentally, a similar project was taking place on an even grander scale in neighbouring Nova Scotia. Tons of granite detonated to carve a rail line through the Halifax peninsula were dumped in the harbour for a massive expansion of shipping terminals, while Imperial Oil established a refinery across the harbour.[38]

Charlottetown never approached the scale of industrial waterfront of Halifax, St. John's, or Saint John. Indeed, in 1962 a member of the Nova Scotia Heritage Trust – fighting to protect a series of waterfront buildings in Halifax – commented ruefully that while historic "Halifax is being torn down ... Charlottetown is the only city where buildings of comparative interest and in any number are still seen."[39] But Charlottetown shared in the same arc of ambition and, on a smaller scale, of development in its waterfront, and the footprint is still visible. As scholars Carola Hein and Dick Schubert have so perceptively observed, "resilience" – while usually an admirable trait – can mean "building on and continuing rather than disrupting historical patterns," rebuilding in the same way and location, locking in practices and infrastructure such as concrete wharves or refineries, based on past investments and larger energy networks. The shoreline becomes, literally, sunk capital, discouraging change.[40] On the shore of the old east marsh, after the gasworks and the coal yards came the electric power plant and Imperial Oil storage tanks – the latter two still very present on the waterfront well into the twenty-first century.[41]

Postscript: Clues to Water

Today, the Charlottetown waterfront presents a comfortable and even encouraging picture of postindustrial adaptation. The early twentieth century bequeathed a handful of structures well-suited to reuse: postcard-ready wooden warehouses and sturdy brick railway shops, stylistically representative of the period, solidly built but technologically superfluous, all a stone's throw from the water. With the closure of the PEI Railway in 1989, the Charlottetown Area Development Corporation led the effort to revitalize the waterfront. Oil tanks were removed to the city's edges. Three railway buildings were converted to serve a variety of tourist offices and attractions. Wharves close to the heart of downtown now house the familiar occupants of a prosperous, leisurely urban waterfront: shops, restaurants, a marina, a landscaped park, and a boardwalk.[42] All of this, green and pleasant, built to a human scale, suits the foot traffic of summer tourists. Indeed, Charlottetown has the most accessible waterfront of any of the cities considered here.

At the same time, though, a large hotel, convention centre, and adjacent parking on square wharves represent the typical corporate, heavier "anchor-attraction" projects that generally have not served aging North American cities well. These are dwarfed by the most recent object of affection in port cities, the cruise ship, a vessel so out of scale to the eighteenth-century grid there is no more perfect metaphor of the Anthropocene, or the epoch wherein humanity has eclipsed natural forces.[43] When a cruise ship docks in port it towers several stories high and reaches several blocks alongside, creating a jarring and unavoidable rupture with the waterfront landscape.

There is little sign of either of the wetlands that framed Charlottetown two hundred years ago. What these two watery spaces share – besides their absence today – is how their stories embody the preferences of Victorian Canada and the patterns of urban history in North America. In both places, the decisions made a century and a half ago, whether for a park or for pavement, have generally determined the shape of the waterfront since. Charlottetown in the twentieth century followed the path set in the nineteenth.

Of the two, Spring Park Brook arguably has fared better. It still runs under the city: you can follow its route in a string of parks and parkettes across the residential neighbourhoods north of the downtown, a tacit and enduring concession to the damp depression of a buried river. It resurfaces alongside the Confederation Trail, its upper reaches protected by a substantial block of green space that includes the federal Experimental Farm (which has held the space for urban farming) and the campus of the University of Prince Edward Island. Government Pond has been shrunk to little more than a wading-pool-sized fragment, with a parking lot filling most of the old pond silhouette. But the cattails are still visible from the much-desired esplanade around Victoria Park and from two of the city's best-known historic properties, Beaconsfield and Government House. Though not yet formally designated, the site of the Bog has become an important landmark for the Island's Black community. Spring Park, though diminished, remains a feature of the city landscape.

In contrast, the land once covered by water to the east is an eclectic combination of industrial remnants over asphalt that suggests little of land *or* water. The picturesque cafés in old brick railway buildings peter out amid gravelled parking lots atop the rest of the old railway lands. But a parking lot is never built on virgin space: the expanses of field, gravel, and asphalt are tell-tale signs of what used to be. There are limits to what can be built over the footing of coal yards or petroleum tanks and the deeper outline of a wetland beneath. Further east, the mouth of what Thomas Wright named Charlottetown Creek is now a combination of the spacious and noncommittal: event space on the old fairgrounds, a Tim Hortons, and an enormous amount of parking – still the stuff of urban periphery to be driven past en route to the Hillsborough Bridge. And as you drive onto the bridge, you pass Irving Oil's petroleum tanks.

In an ironic epilogue to this story, there is currently a proposal for a massive redevelopment of the "Eastern Gateway" that imagines a "city between two parks" – a framing that recalls Charles Morris's plan of 250 years ago. The proposal identifies "a watercourse that is currently channelized in the Eastern Waterfront Park area" and recommends this unnamed creek be naturalized. There is, however, no mention of the area's environmental history.[44] And yet, the paths of these two watercourses are exactly where rising water is expected to penetrate the city furthest by 2100.[45]

Over the course of the twentieth century, many port cities experienced an internal separation, a physical distancing of port from city. The mega-infrastructure for industrial shipping migrated out of the city core; the downtown waterfront was reabsorbed into urban commerce, consumer traffic, and civic marketing. This disassociation of port from city has been aided by a "frayed historical vision" of the waterfront, one that relays a partial or incomplete picture of the city's historic relationship with the water. The reclaimed, marketable waterfront often vaults over the more recent industrial past to return to the age of sail; an age of sail that might be historic, with tall-ship replicas, or contemporary and glamorous, with marinas and yacht slips. The city waterfront proper is dedicated to leisure, festivals, or the condominium gaze; the port, with its massive container cranes and fuel tanks, is conveniently distanced from view.[46] It is easy to forget how much of the physical shoreline is artifice, artifact, and common ancestor – in petroleum and pavement – to both kinds of waterfronts. Like other coastal cities, Charlottetown's *is* a positive story of rehabilitation, of a return of green to blue and people to water, but one that exculpates the city from the Anthropocene story. This is a part of our urban past we need to know. No one wants oil tanks back on the downtown waterfront, but neither should we forget they were there. In a sense, we don't have to. Early twentieth-century infill has left us with early twenty-first century parking lots, which, in their asphalt, have petroleum physically baked into their fabric and which only amplify the heat island effect plaguing our cities.

There are options, though. Arguably the most economically sensible, civically progressive, and climatically responsible action has been the conversion of the PEI Railway line into the Confederation Trail. An olive branch to the surrounding landscape, a tangible contribution to public health, a means of immersion without consumption, the trail is also a quiet commemoration of an older infrastructure, a daily reminder of the transformative effect of anthropogenic fuel and technology. And best of all, for the purposes of this story, it brings us from the head of what would have been Charlottetown Creek to past the head of West Creek – crossing most of the urban watersheds described in this story.[47] As Andrew Hurley has argued, historical education is a necessary partner to environmental remediation. Learning about the history of a landscape as "the product of human decisions that could be undone" encourages people to see restoration – even from under asphalt – as an achievable civic goal. Moreover, he points out, "Because environmental dynamics rarely correspond to neighborhood or political boundaries, their interpretation illuminates the relationships between different strands of the metropolitan fabric."[48] In other words, the urban ties that bind may be more watersheds or urban forests than commuter routes or school districts, which tend to sift out by class and race. This enlivens our sense of connection with each other and with the other creatures that need water in our cities. *This* is a way to live with history – and live well in a coastal city.

4
Tidewater Aligned
Halifax, Nova Scotia

"First Steam Packet From England"

At length we have the gratification of announcing the arrival of a steam vessel direct from Liverpool to Halifax, – the precursor merely of a superb line of steam packets which the Hon. S. Cunard, aided by the British Government, has succeeded in establishing between England and the British provinces in North America ... on her passage up the harbour she was greeted by the cheers of multitudes who assembled by the wharves, and when she was brought to, at the north side of Cunard's wharf, a rush was made at once upon her deck, by the young and old, rich and poor, without distinction.

– Acadian Recorder, *6 June 1840, 3*

Halifax: A View from the Water

In the summer of 1840, merchant Samuel Cunard launched a transatlantic steamship service between the commercial ports of Liverpool, Halifax, and Boston. Arriving at Cunard's wharf on the northern edge of the Halifax waterfront that day in June, the wooden-hulled paddle-wheeler *Unicorn* docked in a space crowded by wooden spars and excited townsfolk. Seventy-five years later, construction began on the Ocean Terminals at the southern end of the city, in a project that would entirely reconfigure that waterfront and the public's relationship with the harbour. No cheering multitudes could rush the new railway lands.

This chapter follows the story of how a city's harbour became a nation's port and the effect of this transformation on the urban fabric. Slate and granite shorelines were remade as docks and wharves, then rail yards, container piers, and pavement. Halifax became a city of superlatives: home to the largest port development project in Canadian history to that point, a construction that remade both peninsula and harbour – only to weather the largest explosion in history before the atomic bomb. And yet, not even the Halifax Explosion could derail or even slow the momentum toward larger, weightier occupants of shipyards, docking, and refinery. Of all the cities considered in this book, Halifax bears the greatest scars of the new energy regime.

Just as Cunard's line promised to collapse time and distance in transatlantic sailing, the building of the Halifax harbour through the nineteenth and early twentieth centuries sought to expand urban waterfront beyond human scale. It permanently altered the human presence at, access to, and traffic on the water, leaving the shore to the shadows cast by shipping containers stacked several stories high.[1] Sections of the city that were a littoral twice over – on the urban fringe for one and between land and water for another – were subsumed and catalogued much more strictly by the mandates of national economic growth, industrial wartime, shipping, and the new fuel of petroleum. Paradoxically, the result was a shoreline dominated by technologies of traffic but absent of people.[2]

A Literary Environmental History

But how can a historian understand the relationship between such different geographies and then situate human experience amid greater-than-human forces? How do we locate people in a city remade by mass construction and mass destruction? Much of the cartographic record in the modern era aligns with a suprahuman scale, in everything from engineering schematics to aerial photography that lets us slip the bonds of earth. But literature brings us back to streetcorners and sidewalks, with voice, colour, and emotion. It helps the historian find people in a landscape increasingly designed to eclipse them. Accordingly, this chapter turns to fiction from three of Canada's best-known writers of the twentieth century: Lucy Maud Montgomery, Hugh MacLennan, and Thomas Raddall – all Maritimers and residents of Halifax in this period. Though canonical figures in the country's literary history, and especially the arc of colony-to-nation, their work has not been read as environmental history. But as we will see, fiction reveals much about the lived experience of the city as it was made and remade. Reading fiction and maps together is also inspired by the broad canopy of the environmental humanities and some experiments in "mapping" literary portraits of places.[3]

Historical writing about Halifax has had a strongly literary cast as well. While there are some accounts of the city by provincial archivists, such as Thomas B. Akins's *History of Halifax City* (1895) and Phyllis Blakeley's *Glimpses of Halifax, 1867–1900* (1949), Raddall's *Halifax: Warden of the North* (1948) has remained the landmark biography: even today, it is "often the one book people have read about the city."[4] (Hopefully, they will now also have read this one.) With a title borrowed from Rudyard Kipling, a novelistic style, and Raddall's special interest in the age of sail, *Warden of the North* has faced little competition from academic historiography, which frequently presented Halifax as a measure of Atlantic Canada's fate in Confederation. In this scholarship, the city was a suppressed metropole, an oft-frustrated measure of regional ambition in the British or Canadian orbits.[5] An illustrated social history appeared in 1999 to mark the city's 250th anniversary, but some of the best research about Halifax has been done by Parks Canada historians and archaeologists at the various historic sites around the city, notably the Halifax Citadel and the Historic Properties.[6] And there is no end of popular history and fiction set in the age of sail and the Halifax Explosion.

Imagine sailing into Halifax harbour today. Your view from the waterline is first of shores of green, the tall pine and rocky beaches of harbour islands and Point Pleasant Park at the end of the peninsula. (Those pines are second- or third-generation regrowth.) The end of the park is marked by the tall cranes angled over the deepest water of the South End container port, built in 1969. Past the cranes you see the older port infrastructure of railway yards, the elevator, and the long, low brick buildings of federal immigration facilities that date to the 1910s and 1920s, but which now draw masses of cruise ships alongside. The all-glass windows of the provincial power authority reflect the light and mark the beginning of the boardwalk. From here the waterfront becomes something more of the city. The crowded eclecticism of the downtown core intersperses wooden wharves, rehabbed wooden warehouse buildings, condominium redevelopment, and office towers. This is the face of Halifax tourism: the perfect balance, it seems, of historic and contemporary.

Sailing up the harbour, small passenger ferries churn across your path to Dartmouth – faster, cheaper, and, it has to be said, more pleasant a commute than the two bridges ahead of you where the harbour narrows. North from here the shoreline is again a swath of broadsides, destroyers and frigates docked at the naval base, and then the enormous assembly buildings of the Irving Shipyard. Turn and look back: from here you have the best glimpse of the Citadel, the remnant drumlin carved into British fortifications three centuries ago, still just visible above the downtown core. Facing out to the harbour mouth you are presented with a composition of green, reforested shore, blue water, and the white of the Imperial refinery at the end of Dartmouth. Set against the expanse of open ocean, it seems an easy coexistence of the civilian, commercial, industrial, and military.

But the view from the harbour reveals another truth about Halifax: most of its waterfront is cordoned off, purpose-defined, and inaccessible to the general public. Pedestrian traffic is confined to the few kilometres of boardwalk running between the anchoring destinations of historic sites at Historic Properties and Pier 21.[7] The routes to the water – through Point Pleasant or along the boardwalk – are urban channels carved out and narrowed as industrial and state occupants expanded their presence on the waterfront. The making of *this* Halifax is the story told in this chapter.

"Functionally Magnificent": Halifax Harbour Through the Nineteenth Century

The town named Halifax was planted in 1749 as a British response to the French fortress at Louisbourg on Cape Breton, which had guarded the Gulf of St. Lawrence since the early eighteenth century.[8] The location of the new town was an ideal spot in geopolitical terms: midway on the Nova Scotia peninsula, a forward bulwark to reassure those on the American seaboard (most vocally, Massachusetts) of Britain's commitment to the defence of its American colonies. But it was the harbour that was the real attraction for a naval power, offering "one of the most secure and commodious havens in the world."[9] The harbour at K'jipuktuk, in Mi'kmaqi, is one of the deepest in the world, and – at least until the era of climate change – partly sheltered from Atlantic storms by a series of low harbour islands and a southeast orientation. The shores were relatively short and inelastic with bedrock but had much more moderate tides than in the Bay of Fundy. A healthy source of fresh water emptied at the southeast edge of the peninsula.

The town's military raison d'être framed its geography. Engineers laid out supporting batteries around the end of the peninsula and at the south end of the town grid. The naval yard was sited protectively in the Narrows,

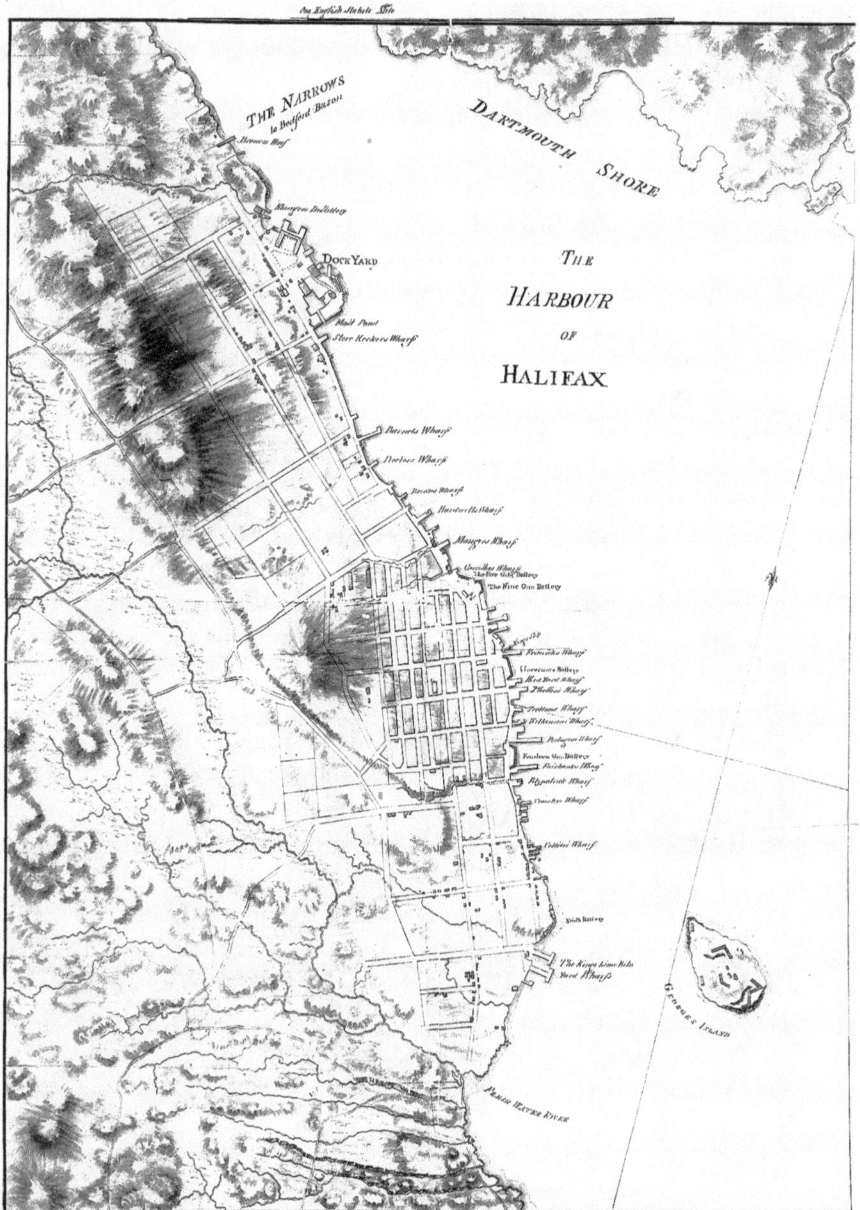

"The harbour of Halifax" (ca. 1777?). Note the wharves interspersed with batteries and fortifications and the number of streams draining into the harbour. | William L. Clements Library, University of Michigan Library Digital Collections closer to the extensive anchorage in Bedford Basin and further from the gales and any attacking forces in the outer harbour.[10] The construction of the naval yard marks the earliest record of an altered shoreline, as sailors filled in breastworks and timber-framed wharves, reinforced the beach with stone and gravel, and began building a harbour wall. Still, Commodore Samuel Hood arrived in 1767 to find much of the yard sodden with poor drainage. Plans of the yard show why: it may have been located strategically with regards to water, but it was less so with regards to land, tucked under a precipitous slope with tendrils of fresh water draining through the yard site into the harbour. Hood reported to the Admiralty that "the greatest part of the yard is made ground, and as large stones were thrown in without being properly filled round, the frost has proved it a kind of subterraneous work."[11] Despite these soggy beginnings, the naval yard would grow to become what one historian judged the largest industrial complex in British North America.[12]

Being a base for the North American squadron was good for business. By the turn of the nineteenth century, the loyal outpost north of New England was prospering from business for the Crown and the profits of privateering (which was, in a sense, also business for the Crown) more than the rest of Nova Scotia.[13] The governor formally acknowledged the burgeoning commercial reality on the civilian waterfront and awarded a market slip in the centre of the town grid "for the use and benefits of the Inhabitants ... to be always reserved and improved as public grounds."[14] While naval traffic would ebb and flow with military deployments, by the 1820s Halifax had developed a healthy trade with ports elsewhere in the Maritimes, the United States, the Caribbean, and even India. The commercial waterfront, between the Dockyard to the north and the Grand Battery to the south, was claimed by an entrepreneurial class who began expanding their shipping and storage capacity by filling in water lots with packed earth and gravel to support longer wharves and larger warehouses. Still, Water Street remained a path "laden with mud and sometimes under water," which, in fact, maps from the 1750s had warned it would be.[15]

But the biggest changes to the shape of the shoreline would be made south of this commercial centre. The original shoreline curved sharply inward, with a promontory offering a good location for a battery facing out to the sea. Midway between this Grand Battery and the fortifications at Point Pleasant was the mouth of Freshwater Brook, the most substantial and thus the only named water course in the city (on the map above it is labelled a river). Freshwater Brook ran the entire spine of the peninsula, collecting several tributaries on its way; the Commons behind Fort George

J.S. Clow, *View of Halifax in Nova Scotia* (1840). | Library and Archives Canada, 1970-188-1985

or the Citadel was sometimes described as a swamp. Just before reaching the harbour, the brook crossed Pleasant Street, a long road that ran along the shore from the edge of the downtown to the pines at Point Pleasant.

J.S. Clow's portrait, finished the same year Cunard launched his Liverpool line, captures "the ambiguity of urban character" of Halifax at this moment.[16] Like most urban portraiture of this period, Clow chooses to situate the city proper at a distance, hoping to screen the grimy reality with a greener foreground. In the painting, we are positioned at the lower end of Pleasant Street looking back at the downtown, much like our imagined arrival by boat earlier in this chapter. Here we are on the edge of Point Pleasant. Stumps in the foreground remind us that the forest had been cleared for military use. (In September 2003, Hurricane Juan levelled much of the park's forest stands, generating widespread grief but supplying something of a history lesson in what the park would have looked like two centuries before.) By Clow's time, Point Pleasant had become the city's exurb park. The road offers a peaceful and presumably respectable promenade along the water's edge; people stroll along the road with space between them, in sharp contrast to the jumble and density of the city in the background. Steele's Pond, a good-sized pool of water adjacent to the tidal shore, makes the shoreline seem even more porous. Men pull nets into a small dory in the inshore shallows and a couple of men-of-war and smaller schooners sail the calm harbour. The industrial corridor that had been established along Freshwater Brook – the familiar mix of tanneries, foundries, and the like making use of the brook as a source of water and a means of disposal – is carefully hidden behind a rise of shoreline, with only a tall smokestack marking the site of the gasworks at the edge of the southern escarpment, still a distinctive edge on the peninsula.[17]

Halifax gradually gathered about itself the trappings of municipal governance through the 1840s, while the colony of Nova Scotia won a form of responsible government – the first in the British empire to do so. As in other nineteenth-century cities, urban crisis often drove urban reconstruction: cholera and gutter waste prompted improvements in water supply and sanitation networks; fires in the downtown core offered the burgeoning merchant and banking sectors an opportunity, in that managerial euphemism, to rebuild more grandly in stone. Rebuilding and the demands of military housing and fortifications spurred mechanization in the building trades, from wind and water power to steam power, making brickworks and foundries "a more industrialized component of the social order and in the physical fabric of the late nineteenth-century city."[18] Freshwater Brook was gradually corralled either in sewers or – for its cleaner upper reaches – in more pleasing ways in the horticultural gardens. Such constrictions also made Halifax and cities like it more impermeable, presaging the environmental concerns of today: rainwater runoff was intensified in tightly channelized waterways, while green spaces that could absorb water and heat alike were turned into stone.[19]

If Halifax never rivalled the metropole of Montreal, by Confederation in 1867 it boasted a full slate of financial institutions; industrial manufacturers; a municipal infrastructure of gas, light, and street cars; and, perhaps most importantly, a railway/steamship complex headquartered at the deep water in the Narrows.[20] Rail lines to Windsor, Truro, and the Pictou coalfields were built through the 1850s and 1860s in the painfully piecemeal and parochial manner of most railway projects in mid-century North America. But Confederation with the Canadas promised a line that would connect the harbour at Halifax to the markets of the St. Lawrence Valley, which were otherwise sealed up with ice in winter. Indeed, Canada's founding constitutional document, the British North America Act, declared that

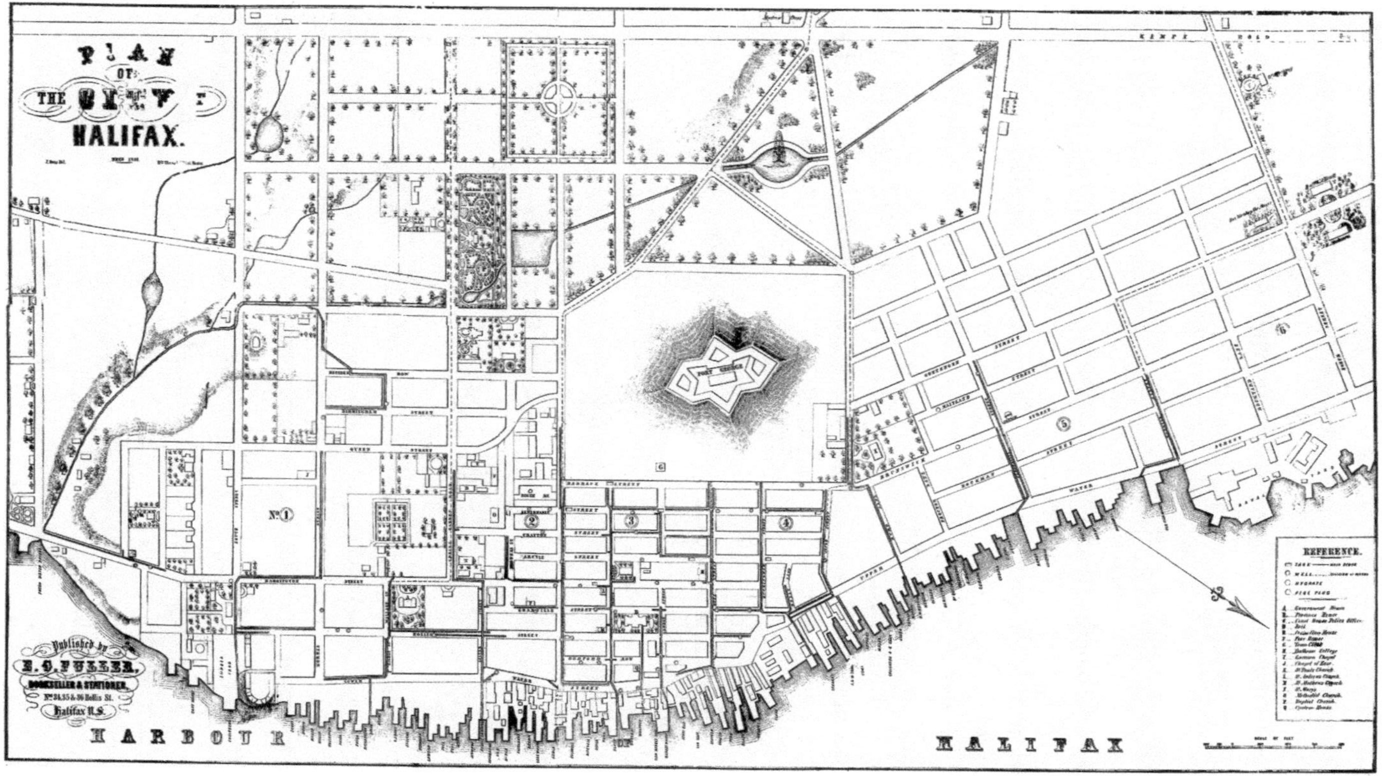

"Plan of the City of Halifax" (1851). We can follow the long course of Freshwater Brook, corralled in ornamental fountains and ponds on the Commons and in the public park, then running around the south edge of the downtown to empty into the harbour. | Halifax Regional Municipal Archives an Intercolonial Railway (ICR) from the St. Lawrence River was "essential to the Consolidation of the Union" and should be completed "with all practicable Speed." Confederation was in large measure a project of grudgingly acknowledged mutual interest between the colonies of British North America, interest telescoped by common anxiety over imperial protections and American acquisitiveness. The majority of Nova Scotians actually opposed the deal and for decades afterwards would complain of Ottawa's preoccupation with westward expansion and federal economic policies that seemed designed primarily to buttress the standing of central Canada. But the debate over Confederation divided Nova Scotia along geo-economic lines. Merchants and coastal communities dependent on international trade and sail generally opposed merger with the inland Canadas, "with which they have no connexion ... being frozen up for five months of the year," while towns through the northern coalfields and along the proposed ICR route were more often in favour.[21] Halifax, though, was both a port town and now a railway depot and stood to benefit as disproportionately from railway traffic as it had from the presence of the Royal Navy.

The ICR was largely assembled from smaller rail lines and, despite its value to eastern Canada, would soon be overshadowed – in both freight and symbolism – by the construction of the Canadian Pacific Railway to Vancouver in the 1880s. Nevertheless, by 1876 trains could run from Quebec to Halifax, with two depots for passengers and cargo at the north end of the downtown core. As a port and railway depot, Halifax was able to successfully marry and layer new transportation infrastructures on top of each other and along its harbour; as Hugh MacLennan wrote, "Halifax was obviously more than its buildings. Its functional aspect was magnificent."[22]

What these interlocking infrastructures had in common now was their fuel. The two traffics converging in Halifax harbour by the 1870s and 1880s were icons of the mineral energy regime. The naval yard had long occupied an important place in the British Atlantic because of its facilities for hull repair; in 1889, the Halifax Graving Dock Company opened the largest dry dock on the North American seaboard – surpassing the one completed at St. John's five years before – in part justified by the fact that Halifax saw more ocean steamships than any Atlantic port after New York and Boston. By now Halifax was also part of a global coaling infrastructure that ensured Britain's global naval dominance.[23] It was an energy entrepôt in the fullest sense. As a coaling station, the terminus of the ICR, and the presumed

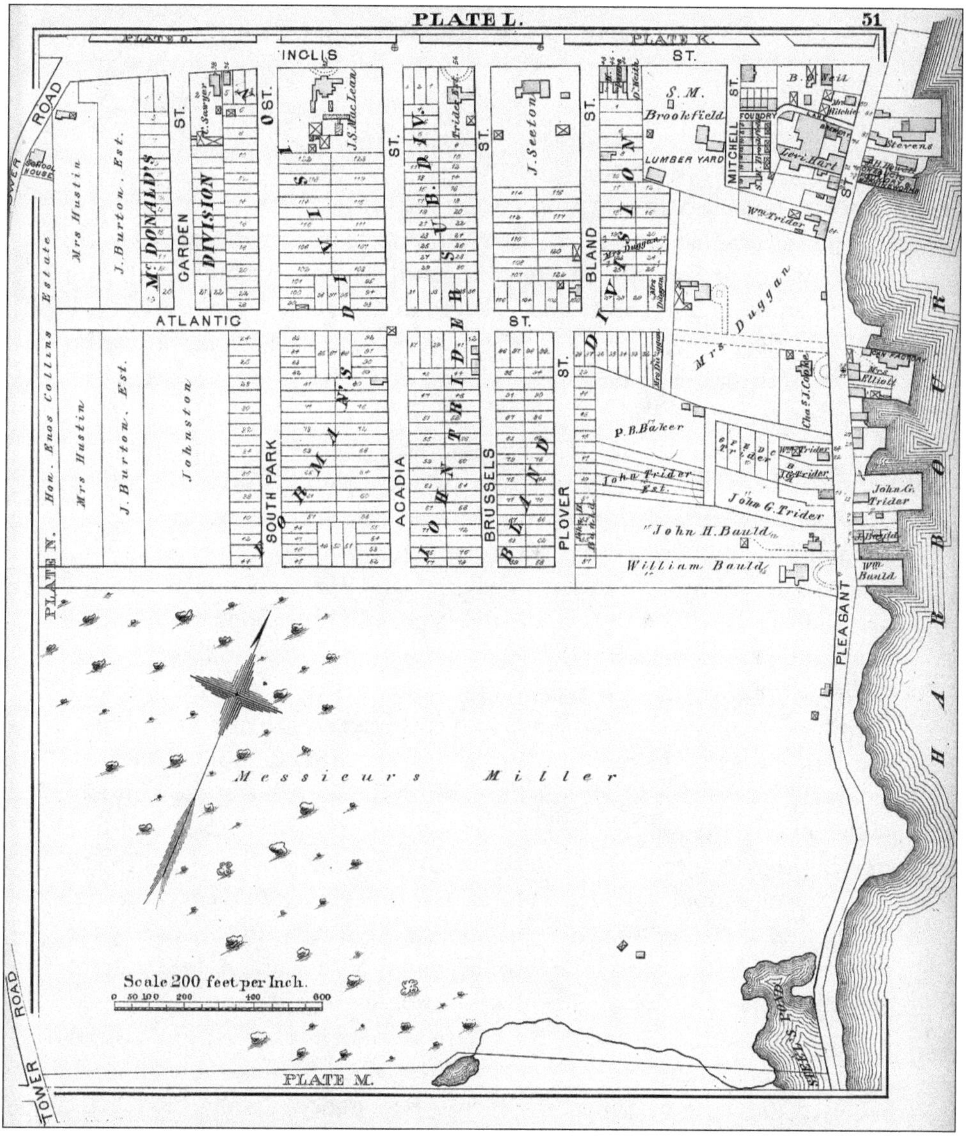

H.W. Hopkins, Plate L, *City Atlas of Halifax, Nova Scotia, from Actual Surveys and Records* (1878). | David Rumsey Map Collection, David Rumsey Map Center, Stanford Libraries gateway to the Dominion, Halifax anticipated a future of prosperity and influence. Such anticipation, powered by the momentum of the new energy age, would transform the peninsula beyond recognition.

A Shoreline Reserve

The other notable change in mid-century was the expansion of new suburbs beyond the original town plan, driven alternately by preferences of wealth and demands of necessity. Fashionable subdivisions carved the south end of the peninsula into estates and villa grounds; tighter housing appeared in the north end, near the dockyards, barracks, and the new railway yards; and the institutions of hospitals, prisons, and asylums were relegated to the urban fringe.[24]

The insurance plan seen in Plate L on page 109 illustrates the changing and varied character of the southern shoreline. While the lower half of the plate, and the Halifax peninsula, remains farm and pasture, the properties along the south end of Pleasant Street are a combination of semirural, residential, and industrial, both occupied and speculative. The curving drives of the carriage class were within sight of the harbour but also uncomfortably or perhaps appropriately close to the sources of their wealth in the various industries – a canning factory, brewery, lumber yard, and foundry – along Freshwater Brook and clustered near the water. This was the corner edge of the mercantile triangle, stretched out along the waterfront. If Clow's painting affirmed the role of the shoreline in the aesthetic preferences of the well-to-do, H.W. Hopkins's insurance atlas established its definition as property within the urban economy.

Freshwater Brook supplied an informal border between the commercial centre and the more spacious suburbs of the South End, between the congestion of the city and the second growth of the park. This section of the waterfront, known as Greenbank or the Freshwater Esplanade, held a transitional space: "One of the prettiest spots *in* Halifax, but a short distance *from* the city," as one guidebook described it.[25] The route, though, required some maintenance. Between spring flooding along Freshwater Brook and erosion from harbour waves, the public beach occasionally needed shoring up, as it was "formed of *made land* with a water wall."[26]

The inshore waters of the harbour in the cove below the curve of the old southern battery (see page 111) attracted respectably genteel activities akin to promenading on Pleasant Street. For the middle and upper classes of Victorian Halifax, it was a popular place for yachting and boating. But

Albert Ruger, *Panoramic View of the City of Halifax* (1879) (detail). | Library of Congress, World Digital Library the unsegregated nature of open water in the harbour meant that less savoury elements of maritime traffic occasionally crowded on the proper and bourgeois. The young folk of the 1895 novel *Miss Dexie: A Romance of the Provinces* regularly picnic and boat on Halifax harbour and up the shore – benign and respectable pastimes. In the novel's most dramatic moment, however, our fair heroine is trapped alone in a small boat. Fleeing a jilted but armed lover and trying to evade a group of drunken sailors "charged with hell-fire in the shape of Water Street whiskey," Dexie is rescued by a British officer's gallantry.[27]

Country in the City: Lucy Maud Montgomery's Portrait of Halifax

The *frisson* of Dexie's near-miss may have been exciting, but it was unusual. In a period when so much of the civic and cartographic record focuses on projects of construction, the literary voice is particularly valuable, as a complement to the material emphasis in urban history and a measure of human sentiment and experience. It allows us to see how people created spaces and meaning within the built environment. In fact, most descriptions presented the waterfront as a respite *from* the material configurations of the city and moral dilemmas of the surge to progress. This was certainly the view popularized by Lucy Maud Montgomery. With the publication of *Anne of Green Gables* in 1908, Montgomery would become one of Canada's best-known writers of the twentieth century.[28] While we generally associate Montgomery with the red roads and apple-blossom orchards of Prince Edward Island, she did have some experience with cities like Charlottetown, Summerside, and Halifax. She spent a year (1895–96) as an undergraduate student at Dalhousie University, and then another year (1901–02) working for the Halifax *Daily Echo*.[29] Though Montgomery knew of the urban streetscape, she focused almost exclusively on parks, gardens, and other sites of Progressive-era beautification in her writing. The city was, for her, redeemed by the nature found in sidewalk cracks.[30] This is why the south-end shoreline and Point Pleasant Park figure so prominently in her descriptions of Halifax, whether in her journal, the *Echo,* or her published fiction.

Unsurprisingly, Montgomery is reconciled to the city when it is most like the country. She was able to find these similarities most frequently at the city's edges and most frequently by the water. Consider her 1901 account of "the glory of an October morning," which concedes that "the early morning is charming, even in the city, when the ends of the streets run off into blue hazes through which the harbor waters cream and sparkle."[31] The harbour supplies an ever-present alternative to, well, the present:

> Last Sunday afternoon I went for a walk in the Park ... just as I would be getting into a rapture over the aroma of the pines or enthusing over a glimpse of the Arm framed in feathery boughs, Polly would say,
>
> "Cynthia, what color do you think I would better select for my new coat – automobile red or bottle green?"
>
> Before us was the silver-gray harbor and on our left the city, scarfed in its mists. Theodosia softly quoted Kipling's lines about "the warden of the honor of the North," as we watched the sun, breaking suddenly from a dark, low-hung cloud behind us, send a flood of splendor over the tossing waters and the purpling hills beyond.[32]

With something approaching literary whiplash, Montgomery suppresses Polly's verbal intrusion with its reminder of the automobile age with older, romantic phrasing, hides the city in harbour mist, and redirects our gaze to the timeless waters around the harbour. Notably, the harbour view is unpeopled, so there is no reference to maritime traffic by which one could date time or technology.

The contrast between city and park was the inspiration for one of Montgomery's earlier short stories, "Miss Cordelia's Accommodation." In this story, a soft-hearted spinster decides to take factory children to the city's outskirts on Saturdays for fresh air, over "the bridge that spanned the river between Pottstown, the sooty little manufacturing village on one side, and Point Pleasant, which was merely a hamlet, on the other" (presumably inspired by the bridge at Freshwater Brook), with the road "trailing off into the real country in a half-hour's walk" (Pleasant Street). The river/brook encircles and contains the city so that the children can reach the "green fields" of health and happiness beyond.[33]

But Montgomery's most detailed portrait of Halifax appears in *Anne of the Island*, the only one of her novels set primarily in Nova Scotia, and the only one set primarily in a city.[34] That said, the novel also gives her a larger stage to indulge her preference for the preindustrial past. Indeed, the landmarks of "Kingsport" (Halifax) are mostly eighteenth-century monuments and cemeteries, "hearkening back to early Colonial days, and wrapped in its ancient atmosphere ... Here and there it sprouts out into modernity, but at heart it is still unspoiled; it is full of curious relics, and haloed by the romance of many legends of the past."[35]

As it happens, Halifax appears most frequently in the novel as one end or the other of Pleasant Street: the cemetery at the head of the street or the park at its foot. In many nineteenth-century cities, cemeteries served as urban parks, a green space for both repose and promenade, privacy and contemplation.[36] This was especially true in Halifax, where the central location of the town's original cemetery, the Old Burial Ground, allowed unchaperoned young women a safe and sheltered outing or an accessible urban park escape "quite undisturbed by the clamor of traffic just beyond." Meanwhile, Anne and her friends live in a rented cottage at the edge of the park, where the grand avenue "petered out into a plain road," "where there is the lure of gray woods and of gray water lapping on the harbour rocks." With its simple and "old-fashioned" air, sheltered by the arms of old pine trees, the cottage softens the crassness of urban modernity with green edgings and humbles the showy display and carefully cultivated self-presentation of the "big lawn-encircled palace of a tobacco king."[37] The university students frequently retreat to the park, which allows Montgomery to focus her gaze (and ours) on her favourite element of the city:

> They all sat down in the little pavilion to watch an autumn sunset of deep red fire and pallid gold. To their left lay Kingsport, its roofs and spires dim in their shroud of violet smoke. To their right lay the harbor, taking on tints of rose and copper as it stretched out into the sunset. Before them the water shimmered, satin smooth and silver gray, and beyond, clean shaven William's Island loomed out of the mist, guarding the town like a sturdy bulldog. Its lighthouse beacon flared through the mist like a baleful star, and was answered by another in the far horizon.[38]

This is a literary version of Clow's oil painting of 1840: from the height and distance of the park at the end of the peninsula, furthest from the downtown core, at a time of day and from a point of view that cloaks the least appealing and most contemporary elements of the city. It was also a version that blended well with other sentimental tributes to the age of sail proliferating in the early twentieth century, whether W.R. MacAskill's photographs of sea-folk and the *Bluenose* or Frederick William Wallace's stories of *Wooden Ships and Iron Men*.[39] The harbour view serves a particular purpose here: since water cannot be built on, it cannot record historical change or human impact. There is no mention of the fortifications on William's (George's) Island, and even the lighthouse is likened to the natural feature of a star. The space between the brook and the park is crucial – exactly the space that will be filled in and rebuilt.

Montgomery's description of Halifax was not inaccurate, just selective and highly romanticized. For her, Halifax was a city by the sea, not a port. She did not visit the Narrows or the city north of the Citadel, which meant she only saw the part of the shoreline that was most accessible to respectable young women and that was reserved for public recreation.[40] By this point some sports and clubs, like canoeing, sailing, and rowing, had begun migrating to the quieter waters of the Northwest Arm, on the other side of the peninsula. But the south waterfront was still a spacious arrangement of suburban streets and open shoreline, in sharp contrast with the congestion of wharves in the commercial centre. Within a half-mile radius of Pleasant and Inglis Streets, the South End boasted clubs for curling, tennis, golf, and yachting: visible landmarks of "the old genteel south end."[41] A tramcar line followed the old esplanade, depositing passengers at the mouth of Freshwater Brook, where people could walk to Greenbank, the bathing pier, Steele's Pond, and Point Pleasant Park. North of the commercial core the waterfront remained the purview of the Navy and the ICR. There was still a little public access to the water in the Narrows, where a yachting club had held a toehold between the navy's graving dock, the sugar refinery, and the ICR freight wharves. This entire array would change, as would many things in Halifax, in 1917.

The Energy of Progress

It may be difficult for us to imagine seeing a harbour tinted in rose and copper through coal smoke, but to Edwardian Canada these were complementary more than incongruous. The crowded waters of prewar Halifax reflected different expressions of capital growth: a seascape of cargo steamers and picnicking sailboats demonstrated, especially to the growing middle class, that a harbour could be a place of work and profit as well as leisure and contemplation.[42] They did, however, represent increasingly divergent energy regimes, with the technology and tradition of wind and oar confined to recreation while increasing amounts of peninsular space and political concern were devoted to the infrastructure of fossil fuels.

Coal was the primary source of energy for domestic, commercial, and industrial users alike. It played a central role in the shape of urban communities in Nova Scotia, as Cape Breton and the Springhill fields made the province Canada's largest producer of coal and steel by the turn of the century, accounting for half of the country's production.[43] The British withdrew from Halifax, their last standing presence in North America, in

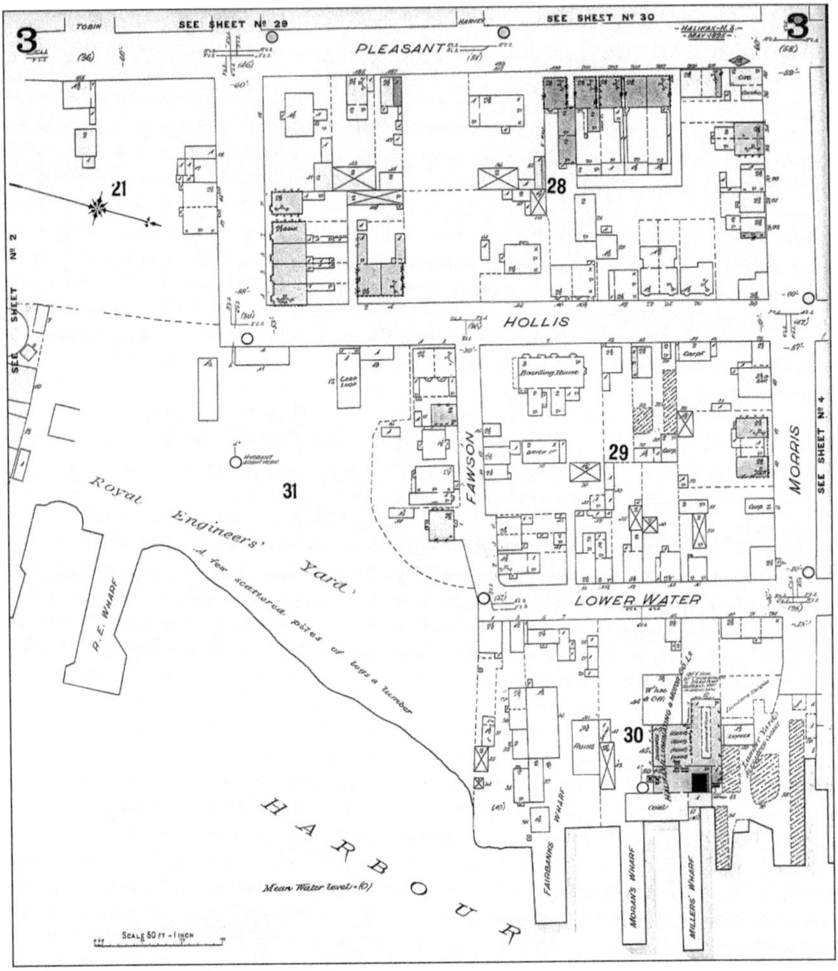

Goad Insurance Map, Halifax, 1895 (sheet 3). | Mount Allison University and British Library Collection

1906, and the Royal Canadian Navy occupied the yard by 1910. The Narrows grew heavier with coal piles for the Deepwater Depot (and passenger station) and the Richmond Depot (and industrial cargo) as the ICR expanded along the shore. The prominent shipping companies – most notably Cunard & Co. – touted their "superior facilities" for coaling ships on site.[44] For residents and smaller businesses, numerous coal dealers could be found through the downtown and near the wharves along Lower Water Street. Suppliers regularly advertised their services in city directories; the 1914 directory, for example, listed at least seven coal and three oil suppliers

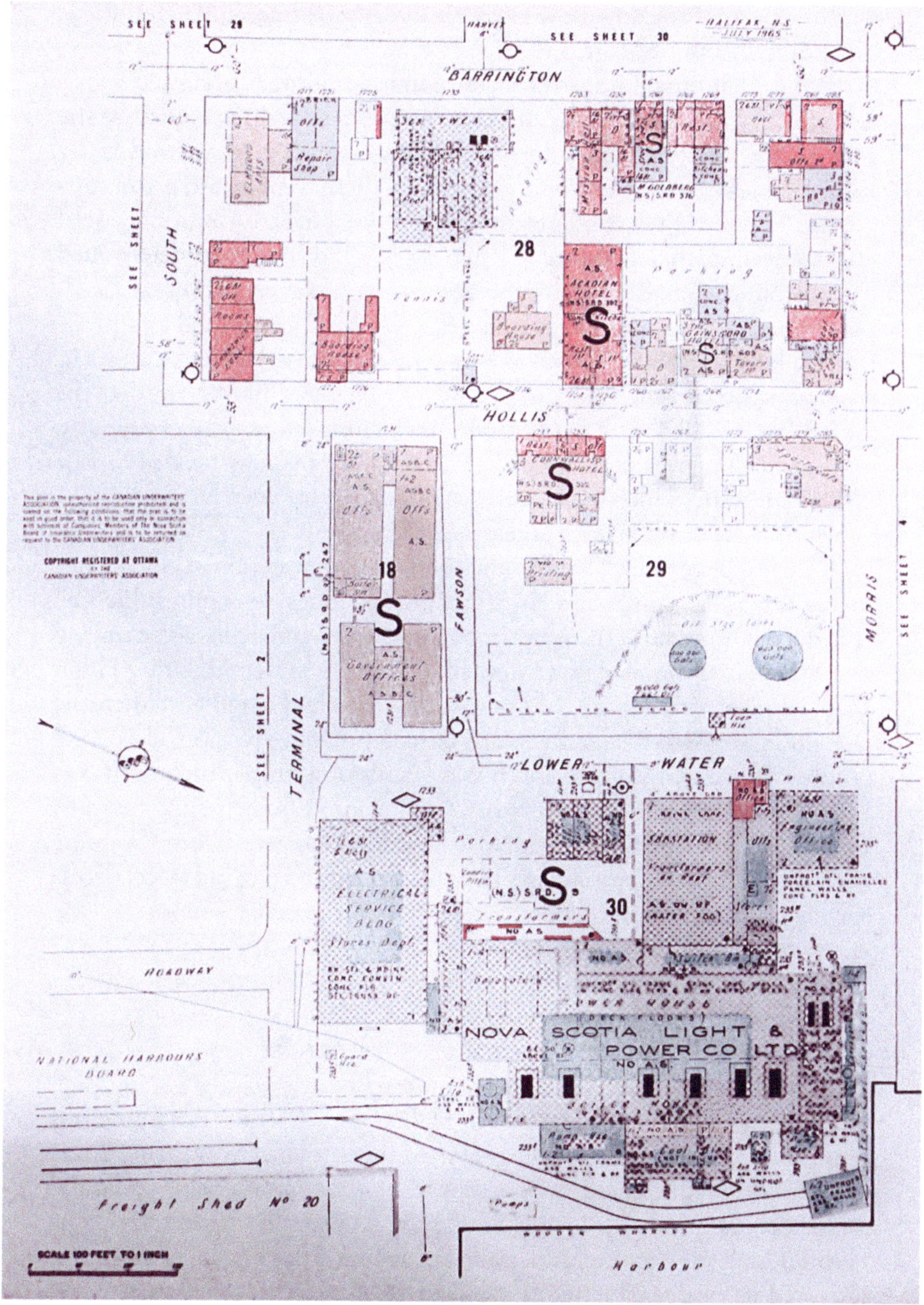

Insurance Plan of the City of Halifax, 1965 (sheet 3). | Canadian Fire Underwriters Association/Underwriters' Survey Bureau. Courtesy Toronto Public Library and SCM/Verisk.

within a block of the downtown waterfront. Meanwhile, Imperial Oil was expanding its bulk storage on the city's outskirts.[45]

The momentum of industrial expansion recast Nova Scotia streetscapes from Sydney to Yarmouth as municipal revenues were directed toward the costly infrastructure needed for industrial services like power and water supply.[46] Heavy, expensive, and networked, these were material commitments that grooved deeply into the city's geography, creating what social scientists call "path dependency": when choices and actions are determined or constrained by older decisions, developing along certain lines. (Such persistence of form and use in place can also be characterized as a problematic kind of "resilience," as we saw in Chapter 3.)[47] The gasworks in Halifax was established next to the Royal Engineers' lumber yard at the Grand Battery site facing the harbour. As the insurance maps on pages 116 and 117 document, the provincial electrical utility is now located on the old gasworks site, which means this single location has been home to power generation for the city since at least 1840.

The new conveniences were not without cost. The combination of flammable cargo, like lumber in large quantities, alongside combustible fuel risked fires throughout the waterfront warehouses and docks – something that other cities in the region knew all too well.[48] As a child, writer Hugh MacLennan witnessed the risks of gas leaks firsthand when his father carried a lit match down into the family basement. The explosion "shook the ground like an earthquake, and it was heard for a mile." Apparently, gas had been leaking in the area for years, seeping into basements and between walls.[49] Gas, coal, and petroleum may have been dangerous, but they were necessary: necessary for a city intent on competing on a world stage, necessary to propel an unprecedented project of coastal construction.

The Gateway to the Dominion: Building the Ocean Terminals

By the turn of the twentieth century, Halifax was churning out numerous publications touting its advantages for commerce and investment. It was not unique in this regard: most cities in North America were doing the same. Halifax, though, could highlight the unparalleled capacity of its harbour, and its geographical position on the Atlantic, as having foreordained its role as gateway to the Dominion. City boosters were quick to enumerate the ways its human assets layered on its natural ones, with

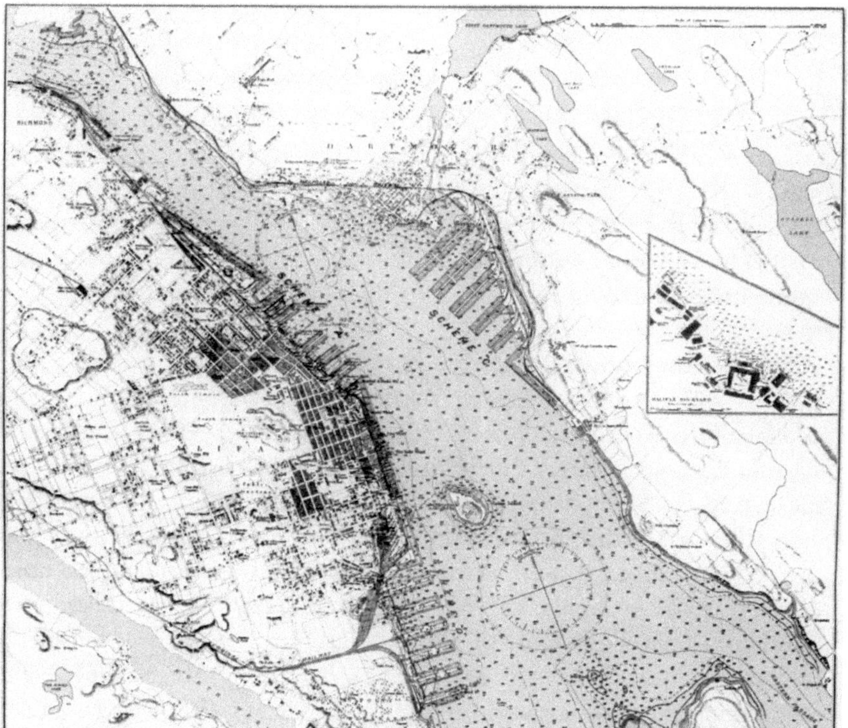

Halifax Harbour | Frederick William Cowie, *Report to the Honourable Frank Cochrane, Minister of Railways and Canals, on Halifax Harbour and the development of a project of modern ocean terminals* (Ottawa, 1913), 39.

"men of energy" and commodious wharves devoted to the quest of progress.[50] As one such guide proclaimed in 1904, "A city by the sea, its people are anxious to avail themselves to the fullest degree of their natural advantages, and every line of business calculated to add to the importance of the place commercially will be fostered ... Halifax has been, is, and must remain the great gateway to the Dominion, the great port of export."[51] Halifax had a steady market in the naval base and rail lines to domestic coalfields. And it had a stretch of conveniently undeveloped shoreline that would invite one of the largest infrastructure projects in Canadian history to that point.

Beneath the bravado, there was growing anxiety about the city's port capacity and its ability to handle the rising levels of immigration traffic to and freight (especially grain) from the western provinces.[52] Ever-larger transatlantic steamers required much more expansive berths, and commercial, industrial, and passenger transports were causing significant congestion along the piers at the railway depot in the Narrows.[53] In 1912, shortly and probably not coincidentally after a Halifax lawyer was elected prime minister, the minister of railways, Francis Cochrane, commissioned Frederick Cowie to undertake a study of Halifax Harbour in order to recommend "a comprehensive Scheme of proposed extensions ... to meet the present requirements of transportation conditions in Canada." Cowie was an engineer familiar with the maritime ambitions of Edwardian Canada, having worked on dredging the St. Lawrence shipping lanes and the port of Montreal. He was also fully committed to the idea that transportation was a national priority. "Can a citizen be loyal to his own city and be equally loyal to his country?" Cowie asked the Montreal-based Engineering Institute of Canada a few years later. (It was a rhetorical question).[54]

The crowds went wild. Or, at least, local supporters of the Conservative government in Ottawa did. "Magnificent Railway and Ocean Terminals that will make a Vastly Greater Halifax" declared the *Halifax Herald* the day after the minister's announcement, making sure to print a congratulatory telegram from the prime minister touting this "WONDERFUL PROGRESS AND DEVELOPMENT." The use of capitals was echoed in the language of scale; the project was described as "gigantic," "mammoth," and "among the world's hives of industry and big enterprise."[55] Former mayor (and ardent Conservative) James C. Mackintosh wrote that it would eliminate the "nuisance" of the esplanade, now rendered "a pestiferous affair" by the sewer outlet. More to the point, "over a mile of absolutely useless shore will be turned into a useful portion of the great transcontinental line," he wrote, "and a majestic station terminal will adorn what is now a receptable of old rubbish." Likewise, the *Evening Mail* praised the "stupendous" project, while scoffing at "the effete south end" for complaining about the loss of mere "picnic grounds" amid "the clank of steam cranes, the whistle of locomotives." The shore as a promenade was of diminishing value compared to its potential as a port.[56]

Cowie took up this theme in his 1913 report, in which he rhapsodized over Halifax's "magnificent natural Terminal Port" – and then promptly recommended demolishing much of it. He identified four potential sites for the proposed Ocean Terminals, including the commercial centre downtown (too costly to contemplate buying out existing waterfront properties) and two sites in central Dartmouth (too removed from most shipping traffic). So Cowie ultimately recommended a site south of the city centre along that "part of the city water front, hitherto but little developed and largely unoccupied," which could be most easily remade.[57] This, of course, was the cove below the embankment, along Pleasant Street to Point Pleasant.

Cowie's report is a remarkable declaration of anthropocentric intentions and its deeply paradoxical assumptions. His proposal for the Ocean Terminals simultaneously trafficked in and disregarded the natural features of the peninsula. The harbour must be made non-natural; liquid would be made solid, curved would be made straight. Fixed infrastructure was required to launch dynamic growth, "in the interest of not only the city but for the improvement of the trade and commerce of the whole Dominion."[58] The unique geography of Halifax was valuable, but in the larger context of transcontinental and transoceanic lines of movement. The human imagination would design a shoreline that diminishes the human presence. And the proposed terminals – like all megaprojects – required capital outlay that made sense only in the speculative quasi-fiction of capital futures.

Visually, the plans for the Ocean Terminals are an exquisite example of an industrial sublime, using the intricate and precise detail of engineering diagrams to simultaneously realize and cloak a project of breathtaking mass.[59] The natural features of Steele's Pond and marshes, the heights of peninsula and depths of inshore water, become a backdrop to an entirely reimagined port of concrete: a fixed, foreign form of human construction in both shape and material. The contrast is accentuated by colour and contour: the blue and grey of the curvilinear shore and the sharp, neat red lines of sheds, piers, and rail lines. Toponymic landmarks like Freshwater Esplanade, Pleasant Street, View Street, and Greenbank were overwritten by an industrial claimant whose only road was, eventually and rather

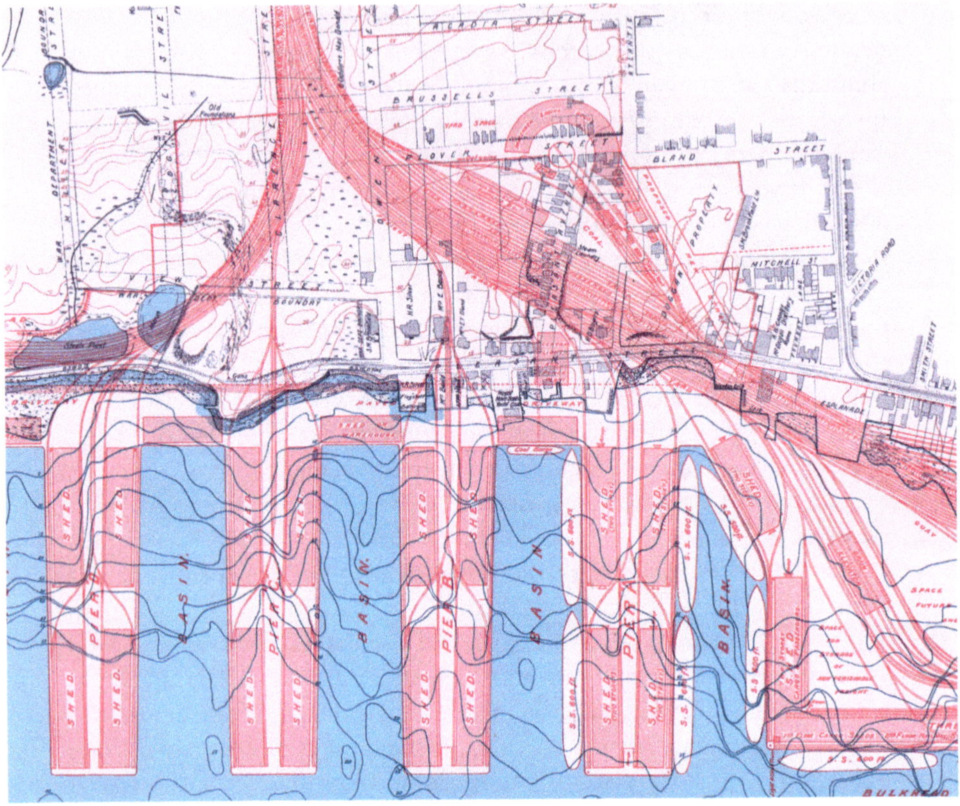

Halifax Ocean Terminals proposed extensions (detail) | Frederick William Cowie, *Report to the Honourable Frank Cochrane, Minister of Railways and Canals, on Halifax Harbour and the development of a project of modern ocean terminals* (Ottawa, 1913), 52.

painfully, named Marginal (a name common in such industrial suburbs). It was an entire planned and anthropogenic ecology designed to circulate freight and passengers.

The construction was, indeed, disruptive to the fabric of the city. Even before excavation began, maps began visually preparing the public for the project and the loss of shoreline. Unlike Cowie's elegantly detailed plans, these maps marked out the terminal lands as a weighty block excised from the rest of the city. In a 1913 brochure, for example, the real estate firm Carrick Limited depicted the terminals as a solid black box, seemingly impenetrable to the public and potentially corrosive to the rest of the city. The text warns would-be home buyers away from the South End, once the "choice residential section of Halifax, but now that the new terminals are to be located there, it no longer bears that distinction,"[60] as though the spatial stain might spread.

When Prime Minister Robert Borden "laid" the foundation stone in October 1915, it meant standing next to a crane dropping a block weighing sixty-two tons into the water. Interestingly, the *Evening Mail* described the event in anthropocentric terms, as "an *Epoch* Event of Civic and National importance." So began months of dredging, blasting out shoreline to depths of sixty feet, and pouring thousands of tons of concrete. Tons of granite detonated to carve a new railway line around the perimeter of the peninsula were dumped in the east-facing harbour. The noise, dust, and reverberation would have seemed incredible.[61] No city in the region had undertaken a project of this scale.

But that was before Halifax blew up.

The Halifax Explosion in *Barometer Rising*

Like almost everything in Halifax, the construction of the Ocean Terminals was, quite literally, derailed by the explosion on the morning of 6 December 1917. The Halifax Explosion was an environmental catastrophe, a product of the petrochemicals of modern warfare. In a harbour crowded with wartime shipping, two ships collided in the Narrows before Bedford Basin. The Norwegian ship *Imo* struck the French *Mont Blanc,* both in port for coaling and loading supplies. The *Mont Blanc,* though, was loaded with benzol and TNT. After the collision, the *Mont Blanc* caught fire. In his 1941 novel *Barometer Rising,* Hugh MacLennan conveys what happened next in a few sentences that capture the terrifying energy of that moment:

> Then a needle of flaming gas, thin as the mast and of a brilliance unbelievably intense, shot through the deck of the Mont Blanc near the funnel and flashed more than two hundred feet toward the sky. The firemen were thrown back and their hoses jumped suddenly out of control and slashed the air with S-shaped designs. There were a few helpless shouts. Then all movement and life about the ship were encompassed in a sound beyond hearing as the Mont Blanc opened up.[62]

It was the most devastating human-made explosion in recorded history before the atomic bomb. Even today, chemical residue from the 1917 explosion is found in sediment in the harbour and on the peninsula.[63]

In 1917, Thomas Raddall was living in a small wooden house on Chebucto Road, in the middle of the peninsula; his school, across the road, would be turned into a morgue for hundreds of victims. He later likened the explosion to a depth charge, but in a description eerily prescient of the atomic bomb: "The smoke of the explosion, a high pillar in the northern sky, was unfolding rapidly at the top in black and white convolutions to form a huge and hideous mushroom, growing incredibly in the sunshine of a winter's day."[64] The combined shockwave and tidal wave destroyed much of the city's North End, home to the naval dockyards, railway lines, shipping wharves, and most of the working-class neighbourhoods. More than seventeen hundred people were killed; hundreds were blinded by flying glass; nine thousand were left homeless by either their houses collapsing or catching fire. The next day, an early blizzard enveloped the city. "Farther on, where the houses stood in silent rows," Raddall wrote, "it was like a city of the dead."[65]

Hugh MacLennan, too, had lived through the event. Born in Glace Bay, Cape Breton, MacLennan grew up in Halifax and attended Dalhousie University before leaving to study at Oxford and Princeton. He was teaching at a private school in Montreal when Canada entered its second world war. With Halifax again mobilized and militarized on an unprecedented scale, MacLennan published his first novel, *Barometer Rising,* in 1941. The story of entwined families during the 1917 Explosion, *Barometer Rising* has been more often read as an allegory for Canada's coming of age, from colony to nation, a theme revived and elaborated in Canadian cultural circles during and after each of the world wars.[66] But what if we read *Barometer Rising* as environmental history, as a snapshot of a city struggling with the growing pains and environmental consequences of the modern era?

In MacLennan's novel, the city's colonial-era plan is being stretched and strained as it absorbs its new role as Canada's eastern port. Geoffrey Wain's merchant enterprise on Water Street represents the older commercial geography, while the new shipyard in the Narrows where his daughter works as a ship designer is the site of wartime industrial activity. The solemn and prosperous Wain house in the South End, with its models of sailing ships, stands in contrast with the close wooden housing for dockworkers in the North End. The tenuous thread of connection between these class-riven neighbourhoods is the harbour, which continues to give the city "its reason for life."[67]

The shipyard is another locus of this tension. Here young Penny Wain negotiates with old men from harbours on the Nova Scotia shore, men who built brigs and barquentines with wooden models and cannot understand either her engineering diagrams or the mandates of mass production. The scale, bulk, and weight of the Anthropocene are most palpable here, as the delicate body of a young woman stands against an "assembly of enormous and potent apparatus": cranes and funnels and coal hoppers. Penny herself questions the reality of these greater-than-human constructions, when "the workmen moving about her looked so small it was hard to realize they could be doing anything that mattered." This surreal experience – the detachment of the waterfront from anything human scale, its absorption into global abstractions – is embodied in a British cruiser sailing into the harbour, as "the incisive outline of the ship seemed to emphasize and sharpen the essential helplessness of all small places in the world to resist the impact of the outer world."[68]

The "floating bomb" that was the *Mont Blanc* left the city "like a field in war-torn Flanders."[69] Halifax was a city built for war but had never been invaded. That morning in December was the first time the city experienced destruction. Like other cultural nationalists of his day, MacLennan saw this as an indication of the precarity of the city and its human-authored superstructures compared to the strength of the rural wild. The explosion reduced much of Halifax to its geological essence, "an oval of rocky soil surrounded by sea and forests."[70] Art commissioned by the Canadian War Memorials Fund in 1918 likewise leaned into the resilience and natural beauty of the harbour and shoreline rather than industrial catastrophe.[71] What if, instead, we consider the Halifax Explosion in a long tradition of industrial disasters, like Springhill, the *Ocean Ranger,* or Lac Mégantic – all accidents born of fossil fuels? The explosion then becomes another "traumatic event embedded in the emergence of twentieth-century energy culture in Canada."[72]

Halifax in 1917 is a jumble of different energy regimes, of streetscapes chaotic with pedestrians, horse-drawn wagons, automobiles, and street railways. In *Barometer Rising,* the Cape Breton labourer Alec MacKenzie cautions his son that studying mathematics will be more valuable than hauling coal.[73] Interestingly, the theme of energy transition and a new petroleum future appears most consistently on the water, telescoped by the question of powering naval fleets. Neil Macrae – who went to engineering school in the "Boston States" – realizes that "now that the navy was burning oil exclusively in its new ships, coal would soon be dropped from all ocean-going craft, and something more utilitarian could be designed if anyone were willing to do it."[74] The harbour is being remade by external interests, specifically the new refinery and what would become an American seaplane base. Geoffrey Wain, having nothing to sell on these terms, instead sells the shore:

> He pointed down the harbour to the place where McNab's Island curves in toward the opposite shore. The American tapped the telescope and set it against his eye, looked for a few moments and passed it on to Eliot. "Is that a refinery over there?" he muttered.
>
> "A subsidiary of Standard Oil. Just seaward are several acres of fields. I think you could rent them for next to nothing if the government were agreeable."[75]

Thus marked the next chapter of Halifax's history: as a petroleum port.

Building a Petroleum Port: *The Nymph and the Lamp*

It is revealing that the disaster of 1917 did not derail enthusiasm for any of the three major building projects that had been announced before that December morning: the Ocean Terminals, the oil refinery, and a shipbuilding facility and dry dock.[76] The McAlpine promotional map (see page 127) from 1920 shows the two harbours of the era: the dense downtown, with commercial wharves interspersed with coal and oil supply, with the all-important transatlantic sailing distances to foreign ports; and the ever-larger industrial properties such as the terminals (in progress but predicted), shipyard, Imperial Oil, and Acadia Sugar fringing outwards from the city centre.[77] The new refinery sat on the Dartmouth shore, directly across the harbour from the Ocean Terminals site and on top of Fort Clarence, one of the eighteenth-century fortifications. Not even a month after the explosion, the *Herald* – making no reference to the recent experience with combustibles afloat in the harbour – proclaimed Halifax the "Greatest Oil Distributing Center." While coal had powered the city's industrialization with steam, oil, and its more streamlined infrastructure of distribution, would eclipse coal after the war.[78] Imperial Oil had been storing and selling kerosene and gasoline via offices on Water Street and depots in the North End for twenty years, and in 1916 began construction on a new refinery at the mouth of the harbour. Though not the first or only oil dealer in the city, by the First World War Imperial emerged as the largest and most influential supplier. Its new terminal and storage depot in Dartmouth were operating by the end of the war and, in a classic

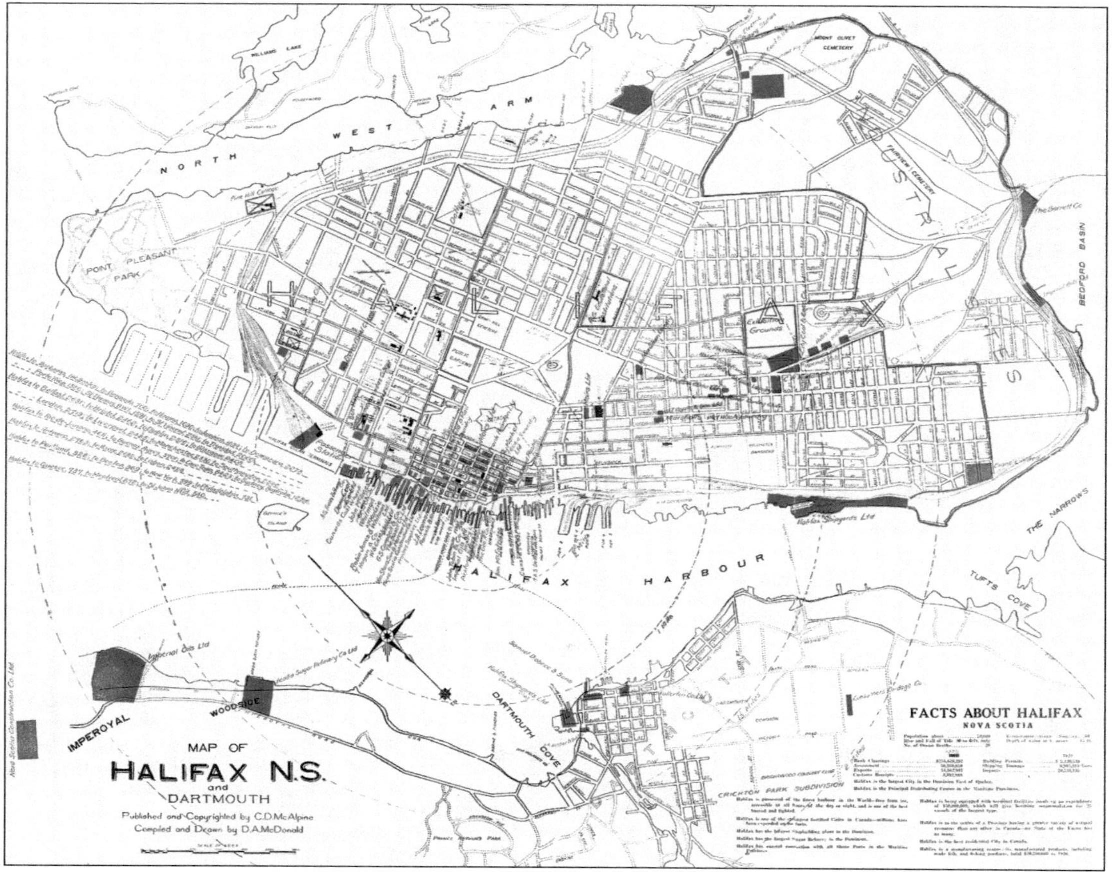

C.D. McAlpine, *Halifax Nova Scotia: The Atlantic Gateway of Canada* (1920). | Nova Scotia Archives Map Collection, V6 239 Halifax, Nova Scotia resource landscape, included workers' cottages in an ordered company town known as Imperoyal.[79] "We have had a coal age and a cement age and it would now appear as if we had entered upon a petroleum age, so vast is the demand and so multiplex the uses for this product of the earth," said the *Herald.*

Even with part of the city wiped to bedrock by the explosion, the war accelerated both coal consumption and the new demands for petroleum. Consider the scene at the beginning of Thomas Raddall's novel *The Nymph and the Lamp* (1950), set in 1920, when Matthew Carney returns to Halifax after ten years on remote Sable Island:

> What the deuce! Even the streets looked queer. All these motorcars! On his way to Marina in 1910 he had counted six on a Halifax street and thought they were marvelous. Now they were everywhere, dodging among the horse traffic, blowing horns, giving off a great stink of gasoline ... What had happened? Was it the war? Or was this "progress"?[80]

By this point Raddall was already one of Canada's more prominent writers, thanks to a series of historical novels set in colonial Nova Scotia and a popular biography of Halifax in *Halifax: Warden of the North*. Perhaps unsurprisingly, given his historical leanings, *The Nymph and the Lamp* is a much more nostalgic depiction than *Barometer Rising*. Based in part on Raddall's own experiences as a telegraph operator on Atlantic ships and Sable Island, the novel is an antimodern romance set alternately on the Atlantic sandbar of Sable, in the tumult of postwar Halifax, and amid the orchards of the still-bucolic Annapolis Valley.[81] The city is the least flattering point on this moral and ecological triangle: a din of typewriters and streetcars, grey and unhealthy, shallow and unfulfilling. Point Pleasant Park plays the same role here as in Montgomery's day, as a safe place of retreat and restoration, especially for single young women. The harbour, too, preserves a glimpse of colour and life beyond the alleys, "like a vertical blue pencil stroke between the buildings."[82] But even Sable Island is no longer safe from either the technologies of war – with steamers and planes overhead and wireless telegraphy – or the energy that drives them. The junior telegraph operator Jim Sargent leaves the Sable station to work aboard a ship bound for Talara: "It's not a very lovely place, I guess. It's in Peru. Oil, you know. It's a regular run. Panama Canal each way."[83]

Meanwhile, the second of the three major projects, the Halifax Shipyards, launched its first ship in 1918. There had been a dry dock (or graving dock) in the Narrows for forty years, but the war made it clear that shipbuilding was in the national interest, and an expanded site was expropriated by the federal government shortly after the explosion. It was then sold to the British Empire Steel Corporation and the Dominion Steel and Coal Corporation. The two conglomerates would dominate Nova Scotia's economic landscape between the wars, during which time the shipyards became the city's largest industrial employer.[84]

And what of the Ocean Terminals? By wrecking the old Intercolonial route into the city from the north, the explosion proved grimly fortuitous for the new rail line being carved around the peninsula. With the old railway station at North Street destroyed, train cars with relief supplies had to travel to the south terminals, even though those were still under construction. Granite excavated from the new rail cut supplied infill for terminal piers, literally rearranging pieces of the peninsula. With sediment and strata considered a marker of the Anthropocene, it is interesting that Raddall used a sedimentary metaphor to describe how the rail yard created its own strata embedded in the city: "The train slowing down in a multitude of tracks that ran off like the delta of a steel Mississippi spreading and flowing to the sea. Isabel could see the masts of ships. Then the final plunge into the vast sooty cavern of the Halifax station."[85] Such trans-dimensional infrastructure would be simultaneously elevated as civic achievement and visibly removed from civic life, literally sunk below street-level: what Ashley Dawson calls the "ubiquity and invisibility" of urban infrastructure.[86]

Improving port facilities at both Halifax and Saint John became a central plank of the Maritime Rights campaign in the 1920s, during which regional advocates sought greater advantages for the Maritime provinces in national transportation policies. The creation of a Halifax Harbour Commission and renewed federal funding helped complete the original Ocean Terminals by the end of the decade, including new immigration facilities.[87] But human contact with this new harbour was minimal. Ships docked alongside a seawall two thousand feet long, a fixed and artificial edge that walled off the water. Historians at Pier 21 National Historic Site have argued that the terminals should be considered primarily a railway enterprise, framed by transportation and commercial imperatives; passenger experience was at best a secondary consideration.[88] The complex was designed for rapid and efficient transfer of cargo and passengers onto trains heading out of the city, away from the water. This was the view for Isabel Jardine sailing out of Halifax at the end of *The Nymph and the Lamp:*

> "Depressing, eh?" said O'Dell's voice at her side. "I like Halifax, mind you, but when you're pulling out on a day like this it looks like the city that God forgot. The war, I suppose. The big explosion of '17 played skittles with the wharves and warehouses, and they were just patched up on a make-do basis ... But all ports look the same, by and large. The citizens see towards the water and everything town-side is fixed up pretty well. Nobody cares a hoot what Jack thinks, coming in at the back door."[89]

Raddall himself was ambivalent about the Ocean Terminals. The rail cut he disdained as an "ugly necessity," but he seemed to admire the scale of accomplishment at the waterfront: "Here the biggest warships and liners in the world could dock with ease, by day or night, at any stage of tide, at any season of the year."[90]

Postscript: "Certain and Inevitable"

In 1956, Halifax City Council hired a professor of planning at the University of Toronto, Gordon Stevenson, to recommend a strategy for urban renewal on the Halifax peninsula. Today his report is associated primarily with the recommendations that endorsed the levelling of Africville, the Black community on Bedford Basin. While most of his attention was focused on the state of Halifax's central and north neighbourhoods, Stevenson found the shabbiness of the downtown waterfront equally distasteful; it was currently "a poor approach to the City" that would be better served by making Water Street a multilane thoroughfare and enclosing the ferry terminal into a multilevel parkade.[91]

Stevenson was writing at a time when the dynamic of petroleum was testing the limits of the peninsula. Two bridges had been built across the Narrows (in 1955 and 1970) to funnel automobile traffic into the city. Additional suburban sprawl and the resulting traffic congestion had sparked proposals for a downtown expressway along the Halifax waterfront in the style of the Spadina Expressway in Toronto, Boulevard Ville-Marie in Montreal, and Harbour Drive in St. John's. Halifax's own "Harbour Drive" project came perilously close to demolishing Water Street but was deflected by community opposition and an intervention from Parks Canada to restore a group of early nineteenth-century warehouses and wharves on Upper Water Street. What became known as the Historic Properties was a pioneering experiment in urban preservation in Canada, and today the

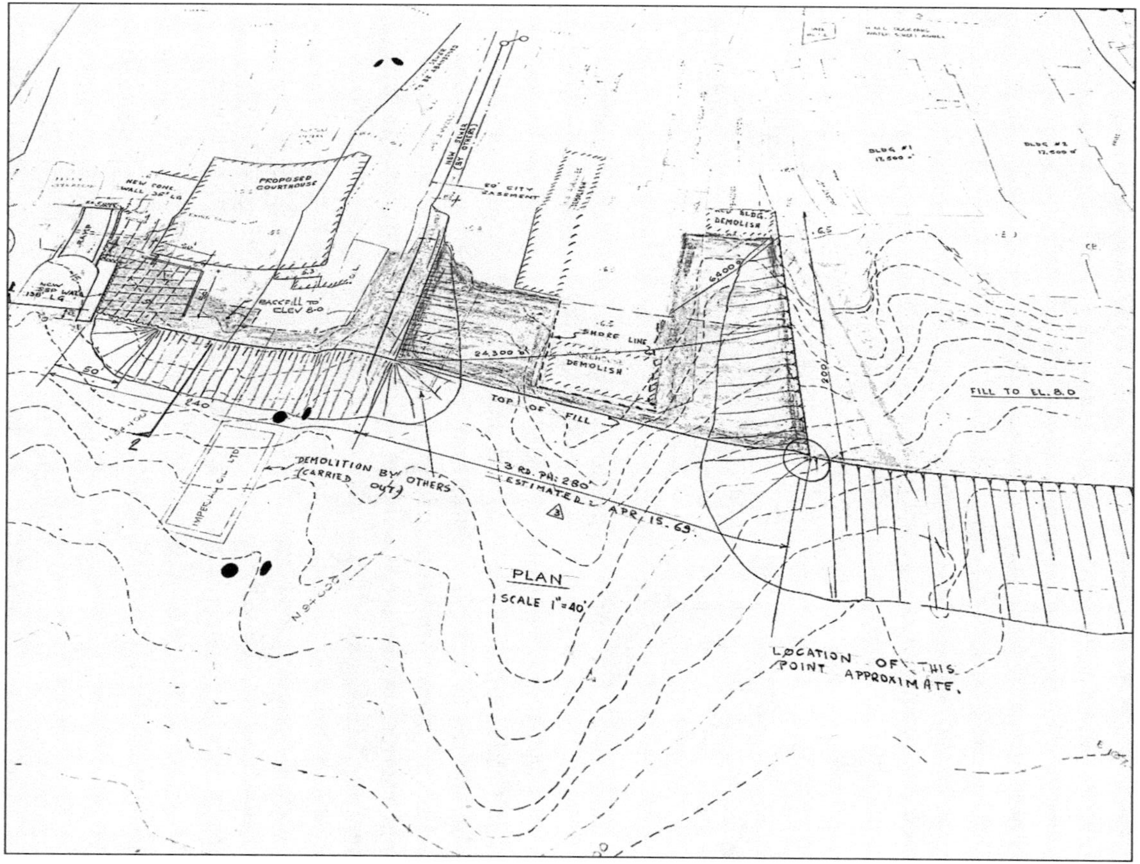

FENCO [Foundation of Canada Engineering Corporation], July 10, 1968: "Land Reclamation at Courthouse Site: General Arrangement." | Halifax Regional Municipal Archives, Engineering and Works Files, Folder Historic Buildings #2 1969-1973 102.39B.1086C buildings are one of the city's best-known landmarks.[92] But even they were not immune to infill.

By the late 1960s the city was knee-deep, literally, in excavation projects in the downtown district. Entire blocks had been levelled for an office/shopping complex at Scotia Square and roadwork such as the Cogswell Interchange. The city had unearthed tons of fill and needed more space for construction. (You probably can guess where this is heading.) The city manager wrote to the mayor and city council in a confidential memo to say that "it appears that the creation of land is the most acceptable and least expensive solution to a very difficult problem which must be solved very quickly."[93] With the proposed Harbour Drive blocked by the new Historic Properties designation, the city was left with an enormous pile of fill at the waterfront, which it tried to dump *around* the historic district. Some of the fill, though, actually ended up under the Historic Properties.[94] Meanwhile, the site of the original 1749 landing at the foot of George Street was now two blocks of fill over the original marshy shore, with a traffic circle for the ferry terminal. City development plans continued to propose infill to expand the waterfront.[95]

But Stevenson had anticipated, and indeed sanctioned, growth along the entire waterfront. As he wrote, "It would seem certain and inevitable that there will continue to be an expansion of facilities in the Naval Dockyard and at H.M.C.S. Stadacona, and that the growth of the port will require additional harbour facilities with new railway arrangements to serve them."[96] Stevenson has been largely proven right: to those that have should be given, and the naval and port facilities have only expanded since his report. CFB Halifax is Canada's largest military base, incorporating the HMC Dockyard, Irving Shipyards, Shearwater, and other properties around the harbour, including most recently a new weapons-testing range in Eastern Passage.[97] Port facilities outside the downtown core have grown exponentially to accommodate container shipping. In the 1960s, the Halterm Port (formally the Halifax International Container Terminal) swallowed more shore south of the Ocean Terminals. At the same time, proposals to dump city waste into Bedford Basin were seen as a convenient solution to two problems – too much waste, too little dockage – but were eclipsed by the construction of an enormous container terminal at Fairview, on the basin, which opened in 1981.[98] These sites have been progressively deepened and expanded, including most recently another extension to the South End container terminal in order to host ultra-class ships. The port authority now refers to the port of Halifax as "Canada's Ultra Atlantic Gateway."

It is, indeed, a landscape of the Anthropocene, an "assembly of enormous and potent apparatus."

Meanwhile, the move toward offshore oil and gas on the Scotian Slope in the late 1970s confirmed Halifax as "the first city of East Coast exploration."[99] Tidy infographics from the provincial government present the seabed as lots available for bid in a centuries-old tradition of the dividing and awarding of property, but on a scale far greater, physically removed, and more abstract.[100] It is a history difficult to confront or to commemorate. Unlike the age of sail, the age of petroleum is at once too valuable, too current, and too distant: most of us will only glimpse refineries and tankers from a distance. No one strolling the picturesque Halifax boardwalk would know that the walk begins on a pile of infill (which had been Cunard's coal wharves), crosses an Imperial Oil depot (at what is now Historic Properties) and an Irving Oil wharf (now Sackville Landing), leads past the old gasworks and a new condominium named The Cunard, and ends at the original Ocean Terminals.[101]

Why does this story matter? We need to recognize not only just how much the nineteenth century shaped our coastal cities, but how deeply entrenched we remain in its development ethos. The province of Nova Scotia, like its counterparts in the federation, remains committed to facilitating fossil fuel extraction offshore. On both Canada's east and west coasts, cities are positioned as energy entrepôts, with pipelines running to tankers. The Irving Shipyard is building patrol ships for the Northwest Passage; something that *could* be right out of the nineteenth century. The language and directives of Cowie's 1913 report, emphasizing use, growth, and "national interest," seem not to have aged a day. In the city centre, a series of larger and bulkier occupants – the Nova Centre, Queen's Marque, various high-end condominium projects – squeeze ever more tightly Raddall's vertical blue strokes of water view.[102] Halifax is wrapped in an anthropogenic atmosphere far more than an ancient one.

POSTSCRIPT

Inheritance and Imagination

We have to learn to stay within our boundary lines. It does not take much imagination if all we have to do to develop our waterfront is infill.

– Clarence Spencer, Environment Canada (2001)[1]

As I write this, the Halifax Regional Municipality is debating infill projects at several locations around the city, including Dartmouth Cove, the Northwest Arm, Mill Cove in Bedford, and the original Ocean Terminals. The purported reasons range from increased docking for cruise-ship traffic to expanding the waterfront for private properties. But if cruise ships and condominiums belong to the twenty-first century waterfront, our approach to that waterfront in many ways still belongs to the nineteenth. It seems that we may need to exercise more imagination in our coastal cities.

Imagination brought us here, after all – at least, certain kinds of imagination. The promise of the coast helped determine the position and purpose of these cities from the eighteenth century. Imperial ambitions for global power were succeeded by national dreams of growth, while the cities themselves aspired to unique influence and standing, whether as conservators of the water or gateways to the Dominion. A development ethos that privileged extraction and expansion threads through the age of sail and the age of steam to the present. The urban waterfront is a multichaptered story of artifice and intervention, with many of the same or similar uses grooved deeply in place over time. It is thus a palimpsest, a record of past use but with traces of its earlier form.[2] The pleasant confines of the boardwalk or harbour trail give us a platform to contemplate the cumulative generations "of enormous and potent apparatus," from the nineteenth-century rail yard to the twentieth-century tank-farm to the twenty-first century cruise ship and condominium building towering over the wooden wharf. We can see the progression of generations of industrial infrastructure and our constructed dependence on fossil fuels. The edges of water, fresh and salt, supply an archive and a vocabulary for understanding this larger history.

It also allows us to see (if we choose to) why our coastal cities are increasingly vulnerable. Past choices to reshape the shoreline in certain ways have built us into new precarities. Our anthroscapes are rapidly proving they are not fortresses against environmental forces. Fire and flood newly animated by climate change are creating images that recall biblical prophecies of catastrophe. As the boardwalks are washed by storm waves and basements are flooded with seawater, we are learning that what George Perkins Marsh called "land artificially won from the waters" is not as secure a possession as those in the nineteenth century had hoped. The seeming permanence of pavement belies our vulnerability in an era of environmental turbulence. Cities are confronting the consequences of so many decades of defying the limits of the littoral – making decisions that are then reaffirmed by subsequent generations. Municipal and provincial climate action plans calling for "coastal preparedness" need to acknowledge that these places were built according to a logic that does not function in this climate era; they cannot be easily retrofitted with a different mandate of adaptation. Hardening the shore has arguably limited our imagination and narrowed our capacity to respond.

But could we *try* to imagine something different? Is there another way?

The narrative of urban growth and nation-building, with all the attendant effects and costs, is sometimes framed as linear, teleological, and inevitable. These are, after all, places where we live today. But what of places that became something different than centres of industry and ports of call? What happened to places laid out along the coast with the same intent for imperial and industrial expansion but never became major cities? What about towns planned for growth ... that never grew? Atlantic Canada holds so many stories of coastal settlement, across centuries, on different scales – is there scope for the imagination here?[3]

Historians are often wary of hypotheticals – the "what if?" line of questions that speculate on alternative outcomes or the paths not taken. There are too many variables, too many unknowns, for us to make confident predictions about the might-have-beens. But in Atlantic Canada, we actually have communities that did *not* become like the cities explored in this book; towns that were intended for the same futures, but followed different paths. Perhaps now we could visit three such places and ask why. Like Halifax, Charlottetown, and Saint John, these towns were sited for settlement with imperial confidence in the mid-eighteenth century and saw commercial prosperity in the nineteenth. Each embraced the language of progress and improvement and aspired to growth, but each navigated the transitions from wind and wood in different ways. Today, one is quiet, economically depressed, and strikingly vacant (Georgetown, Prince Edward Island); one is an out-of-the-way resort town (Saint Andrews, New Brunswick); and one is a World Heritage Site overrun by summer tourists (Lunenburg, Nova Scotia). None of them are substantially different, in size or character, than their colonial footprints. Indeed, these towns are more conventional (and marketable) representations of Atlantic Canada *because* of their size and streetscapes, their saltbox houses and wooden wharves. Most importantly, perhaps, we can imagine, and in some places actually stand on, the older coastline – there is nothing on the scale of Courtenay Bay or the Ocean Terminals or the St. John's Dockyard keeping us from the water.

Georgetown, Saint Andrews, and Lunenburg were all laid out in response to the same concerns of imperial Britain, in almost identical ways, in remarkably similar locations. Each was "planted" on a small peninsula with the town facing in a sheltered southerly direction. Each has retained the grid of the model town, with a central square, aristocratically inflected street names, and wide streets; each embodies the Enlightenment confidence in survey and order. Indeed, all three were laid out by the same Charles Morris, surveyor general of Nova Scotia, as near-copies of his towns at Halifax and Charlottetown; Georgetown and Saint Andrews were also designated as shire towns, or county seats.[4]

Each saw something of a "golden age" in the era of wooden spars and sailing ships, and each still has remnants of a working waterfront, but none became lasting, expansive centres of industry. None lost their harbours to swaths of pavement, cordoned off by heavy machinery and wire fencing. Each was targeted by mid-century experiments in state-led regional development and various private investments, but despite infusions of cash, none was revived as a major industrial site. Each was sited in ways that made more sense for living in a coastal world, but became more inconvenient to reach in the era of fossil fuels. Accordingly, each now is arguably a far more desirable place to *live*. What happened? What contained their growth? Are they any better placed to weather sea-level rise and the other effects of climate change? Can we replicate any of these experiences elsewhere? What can we learn from them?

Lunenburg

The town of Lunenburg is an artifact of a different era of coastal anxiety. Worried about its hold on the formerly-and-still French Acadia amid the ongoing wars of the eighteenth century, Britain's Board of Trade and Plantations concentrated new energies on securing a loyal Nova Scotia. This included recruiting immigrants from the United Kingdom and the Thirteen Colonies to the most arable spots on the peninsula, concentrated along the Fundy shore and Annapolis Valley, and the *dérangement* of the resident Acadians. Such relocations and displacements, voluntary or otherwise, were part of the underlying global expansion of capital and imperial power.[5]

Among the handful of safe harbors on Nova Scotia's Atlantic coast was Merliguesche, home to a small Acadian community, which was "only fit for small vessels, but has the best improvable land of any other along this coast."[6] It was also reasonably close (ninety kilometres) to the new Citadel and military forces at Halifax. In the summer of 1753, the British military resettled here about fifteen hundred émigrés originally from southwestern Germany and western Switzerland on a townsite appropriately renamed the Germanic Lunenburg. (Ironically, even with its defensive perimeter, Lunenburg remained relatively precarious for another century – in part because of its vulnerability to attacks from the sea, especially American privateers.)

Within the larger agenda of imperial security, Lunenburg itself was meant to be an agricultural settlement. Each of the five hundred town lots were assigned rights to commons and farmland beyond (as we saw at Charlottetown), and the original settlers were to be supplied tools for planting. The township grant issued in 1784 acknowledged the work already "carried on progressively by laying out & Locating Plantations to the Inhabitants as they advanced in Cultivation & Improvement."[7] While ostensibly more arable than the Halifax peninsula, settlers at Lunenburg still had to rely on foodstuffs and other provisions supplied by colonial officials for decades after. By the end of the century, it was clear that the town had evolved a different economic profile than originally intended, one that trafficked as much in exporting fish and lumber as in produce:

> Those who live in the Town have amassed fortunes by supplying the Halifax Market with livestock and vegetables of kinds, Barley, Oats, Hay, &c. for which purpose they have near twenty sail of small vessels; the growth of timber is nearly the same as the adjacent settlements. There are no Saw Mills

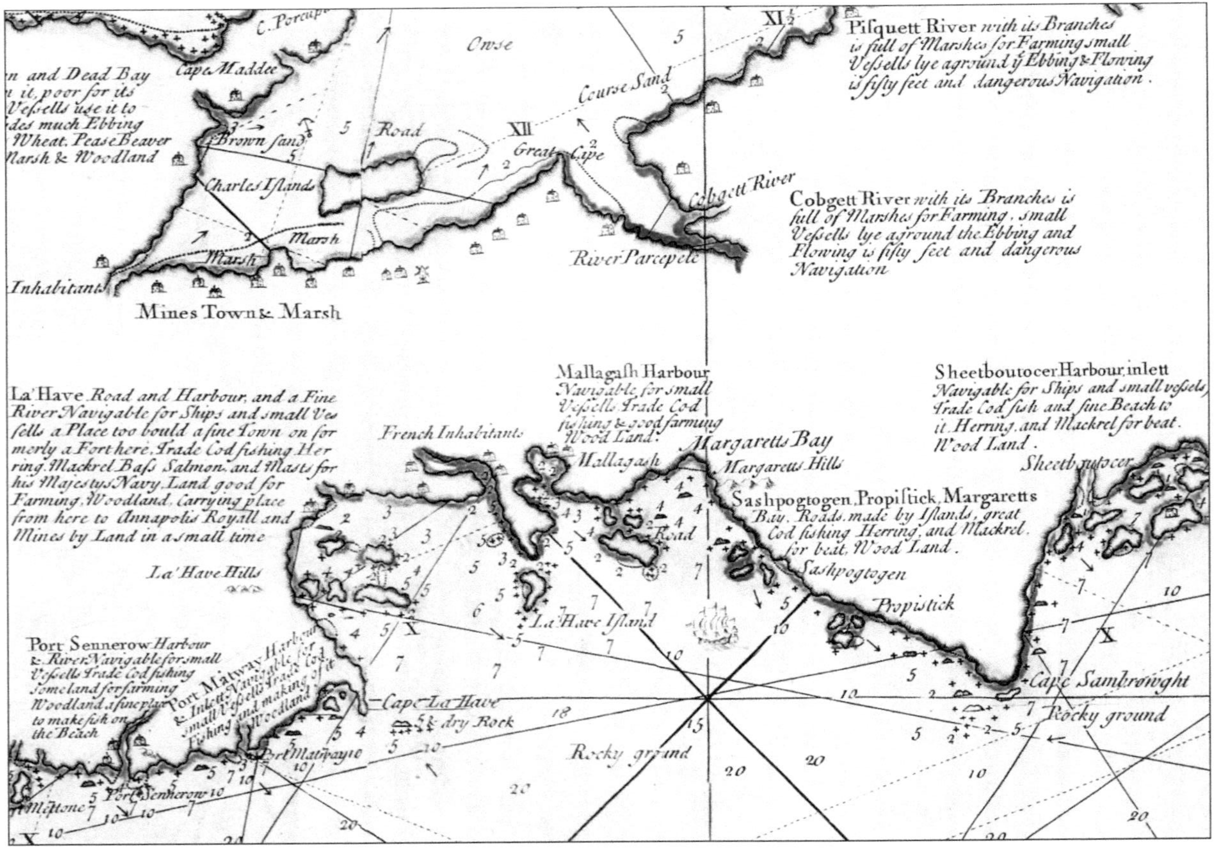

Cyprian Southack, *An actual survey of the sea coast from New York to the I. Cape Briton* (1758) (detail). The map describes Mallagash as having "French inhabitants," but promises it is "Navigable for small vessels, trade code fishing & good farming, wood Land." | Norman B. Leventhal Map and Education Center at the Boston Public Library

> very near the town but at the back of the Township fronting on a part of Mahone Bay, are several, which have been erected some years, and from thence is exported to Halifax considerable quantities of Lumber.[8]

In his influential *The American Geography* (1789), Jedidiah Morse – already predisposed to disparage the British territories remaining in North America on behalf of the new republic – noted that Nova Scotians "subsist principally by the lumber trade, which is supplied by their forests; and by the fishery, which is very profitable."[9]

Indeed, it would only become more so. By the decade of Confederation, the town had emerged as one of the most important centres of the offshore fishery, reaping wealth from building and outfitting the ships (known as "saltbankers") that sailed to the world's richest fishing grounds along the Grand Banks of Newfoundland and the Scotian Slope. Unlike Georgetown and Saint Andrews, which kept their original plats, Lunenburg laid out an annex west of the eighteenth-century grid intended as a fashionable suburb with High Victorian manors, though the so-called New Town would be the only substantial addition to the original plan. By this point Lunenburg was "regarded as the most wealthy and beautiful town on the South shore."[10] In 1921, local shipbuilders Smith and Rhuland launched the *Bluenose,* a racing schooner whose success against American boats made her a provincial and then national icon. With the *Bluenose,* Lunenburg had successfully trademarked the waterfront, and in so doing crafted a kind of economic insurance that would eventually pay out.

But despite its assets in the architectural stock and the silhouette of the *Bluenose,* Lunenburg – like other small coastal communities – suffered a decline in the mid-twentieth century as North America tilted terrestrial. Indeed, as early as 1907 one observer cautioned, "Although Lunenburg has attained an enviable position as a shipping and fishing centre, it is no doubt a fact that its interests are too much bound up in the one industry to assure continuous prosperity."[11] The waterfront was largely turned over to industrial fish processing, shipbuilding, and a local foundry – all of which are still present, incidentally – but many of the larger, service functions drifted to the railway and highway hub in nearby Bridgewater. Lunenburg's population plateaued at about three thousand residents.[12] Like many other ostensibly declining rural regions, the provincial Crown corporation Industrial Estates Limited and federal agencies offered development incentives to try to attract outside corporations to set up shop in lieu of older businesses and proposed an industrial park on the outskirts of town.[13]

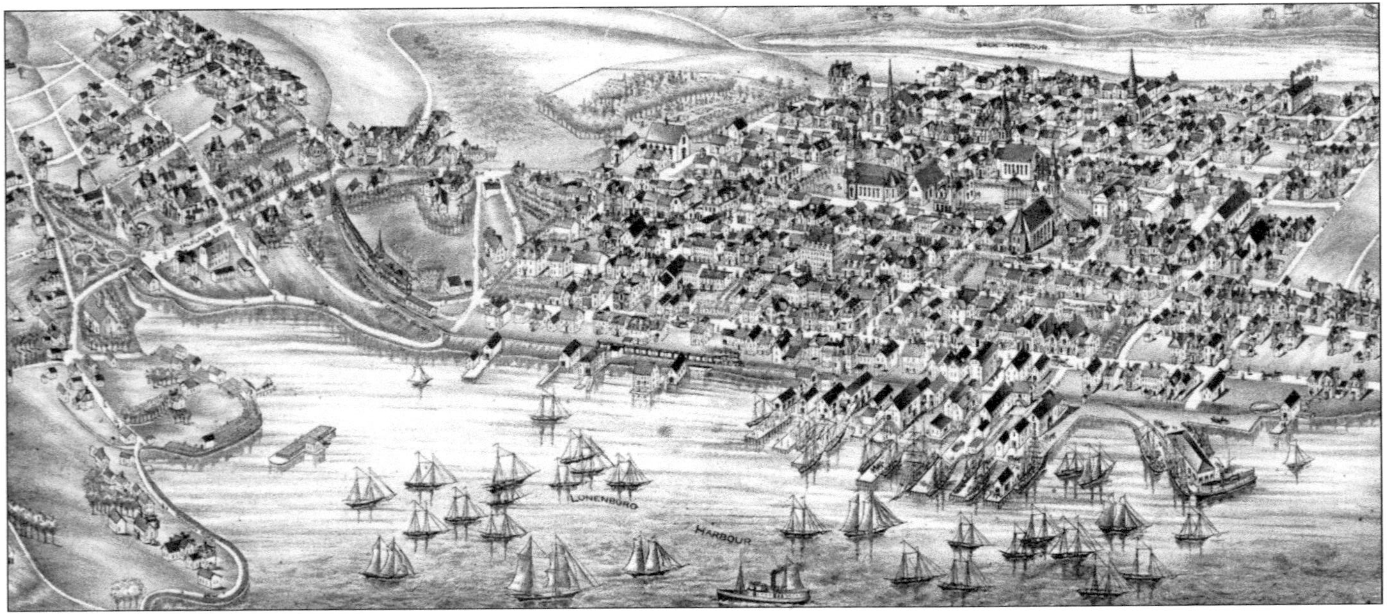

D.D. Currie, *Lunenburg, Nova Scotia* (1890) (detail). Note the line of railway running closely between Water Street and the harbour, with little embankment and a relatively slim profile. | Norman B. Leventhal Map and Education Center at the Boston Public Library

In 1966, the town commissioned a study from Canadian-British Engineering, with Norman Pearson, one of mid-century Canada's most influential planners. (The firm also undertook studies of several other Nova Scotia towns, including Halifax's sister city Dartmouth, the agricultural centre of Windsor, and the coal towns of Glace Bay and Springhill.) The report did *not* take up the cause of industrial parks and investment incentives. Instead, it declared Lunenburg to be "the urban jewel of Canada: a town, still in the human scale, which is the very essence of the Maritime civilization." At a time when many cities, including Halifax, were taking drastic measures in the name of "urban renewal," the report found in Lunenburg no "derelict" space, only evidence of "the ability of Canadians to build effective civilized and beautiful towns." The authors recognized that the town had not urbanized as such, insofar as it had not grown from a rural community, but instead that "Lunenburg had urban values from the beginning."

The key, they felt, was the town's fidelity to its original plan, which stood in contrast to the "speculative sprawls" of postwar North America.

"Town of Lunenburg," Nova Scotia Information Service (1966). | Nova Scotia Archives NSIS no. 19575

As they wrote, "life has a harmony and a scale of contact with the sea and the land behind which is missing in our larger cities and should be nurtured here." Indeed, it was the multiple *levels* of scale that made the town so appealing: the juxtaposition of the "enclosure and intimacy" of the walkable streetscapes fronting the vast panorama of the sea. Lunenburg had not walled off its waterfront; living there meant being daily aware of the water: "Despite the declivities [sic] of its streets and the inconstancy of marine weather it is a place where people walk and where the scene all around them is beautiful, where the views at every corner (both the street pictures and the views out over the harbour) are memorable." The report cautioned against parking lots, but instead recommended that designating space for historic sites, parks, and a greenbelt ("a new use of the old Commons tradition") would serve to keep the shoreline "green and open." As the authors concluded, "Lunenburg is a town in which every street picture gives the evidence of the quality of life in such an environment. It is an inspiration for our generation in creating new urban forms."[14]

To be fair, Lunenburg did have other advantages, chief among them the most famous sailing ship in Canadian history. There was also – as in many Maritime communities by the centennial decade – a growing presence of both the federal and provincial governments on the waterfront and substantial interest in harnessing nostalgia and tourism for revenue. The construction of the replica *Bluenose II* was a corporate and civic investment that supplied Nova Scotia with a "sailing ambassador."[15] To commemorate Canada's centennial, the town opened a fisheries museum that included the last salt-banker still floating at the docks. In 1975, the museum moved to a former fish processing plant fronting the harbour, and the next year Parks Canada established an exhibit as part of a federal commemoration of the historic significance of the banks fishery. The museum was folded first into the provincial and then the national museum network, giving it access to new funding, and renamed the Fisheries Museum of the Atlantic.

But it was the closing of the groundfishery – a five-hundred-year-old harvest and way of life in the North Atlantic – that created present-day Lunenburg as a new version of its old self. Despite the romantic image of schooner fleets under sail, it is important to remember that this fishery "was always industrial, always resolutely global and always had an ecological cost."[16] That cost became painfully apparent by the early 1990s, and in 1992, the federal government closed the offshore fishery as cod stocks collapsed. It was precisely at this moment that Canada (led by Parks Canada) accelerated the campaign to designate Lunenburg a national historic site and World Heritage Site. In 1995, UNESCO recognized the town as "an outstanding example of the planned European colonial settlement in North America, in terms both of its conception and its remarkable level of conservation." (While the overwhelming majority of the town's buildings actually date to the Victorian era, the eighteenth-century street grid is mostly intact.)[17] Notably, the UNESCO designation also acknowledged the vulnerability of the fishery and the way of life that had depended on it. The heritage designation is a silent commemoration of an ecological disaster.

Today, the town of Lunenburg has a complicated relationship with its historical identity. Eighteenth-century streets of "enclosure and intimacy" do not easily facilitate cars or the numbers of global tourists. Lunenburg is one of the most iconic destinations in Atlantic Canada – arguably the country as a whole – and struggles to accommodate the hundreds of thousands of visitors drawn to the waterfront each year.[18] Smaller industrial tenants (mostly ship repair and fishing supply) still bookend the public waterfront space, but the town is caught between questions of affordability, on the one hand, and revenue, on the other. As it turns out, being picturesque is expensive, to both live in and pay for. In 2024, the town council approved a housing project on Blockhouse Hill, just outside the borders of the UNESCO designation (in a project known as "Out of Sight"), hoping to add space for over seven hundred more residents – and their property tax revenue. The next year, however, the council withdrew its approval, concerned about the potential strain on the town's energy grid.[19]

Georgetown

Georgetown was, similarly, a creation of the surveyor's preference for order and symmetry set within the empire's desire for possession. As we know from Chapter 3, Samuel Holland surveyed the Island of St. John in 1764–65, divided the island into three counties, and assigned each a county seat and royalty. For the county seats, he was careful to select some of the best harbours on the island. For the easternmost town, Holland selected a peninsula carved by the Cardigan, Montague, and Brunell Rivers, which French settlers had called Trois-Rivières and which Holland renamed Cardigan Point. Here, he imagined "may be erected a Pier with great ease, and at a small expense, where goods could be shipped and unshipped with great facility and convenience."[20] For King's County he suggested, appropriately, a seat named for the reigning monarch.

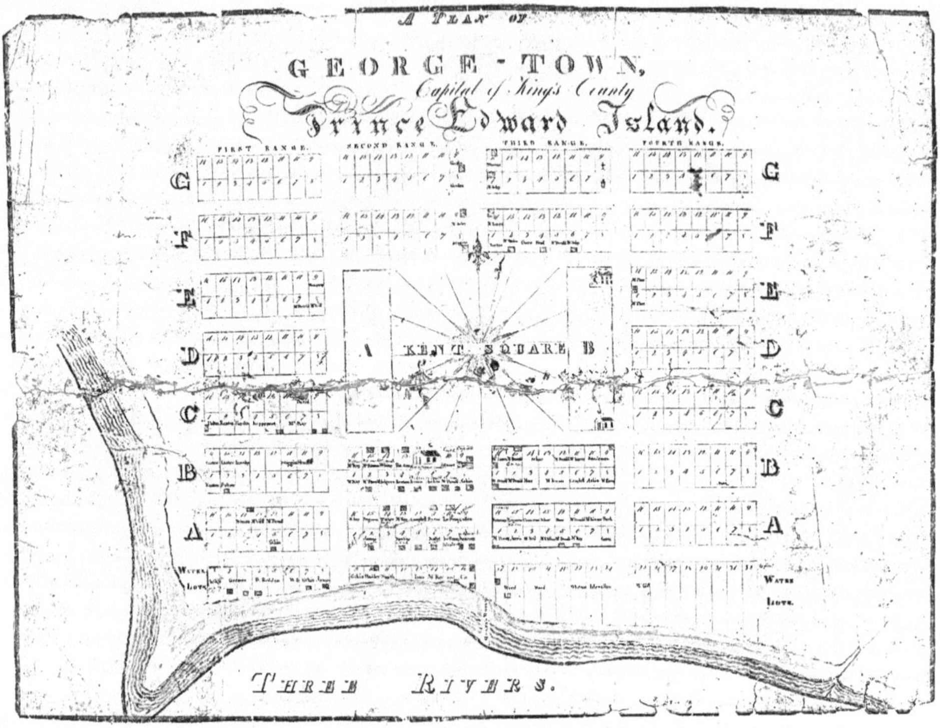

"Plan of George-Town: Capital of Kings County, Prince Edward Island" (1835). | Island Imagined Map Collection, Robertson Library, UPEI imagined: 209313

The ever-present Charles Morris supplied a town plan in 1768, framed on two sides by water, with a central square. Nearly forty years later, John Stewart agreed that the town's "situation is very fine, and the Harbour one of the best in North America." Indeed, Stewart felt that Georgetown had a better location than that of the capital at Charlottetown, "being much nearer the ocean and of much easier access" with a "situation large enough to accommodate an extensive commerce." And yet, Stewart reported that even with such potential, Georgetown "is yet as a Town but in embrio [sic], there being but a few lots granted, and only three or four built Upon."[21] Unlike Lunenburg (or Halifax) there was no initial deposit of settlers; they built it, but no one came. An 1827 census counted 156 residents in the entire royalty – less than a tenth of Charlottetown's population.[22]

This would puzzle observers through the first half of the nineteenth century: Why did such an inviting location, with a harbour that did not freeze over and was easily navigable from the Gulf, remain undeveloped?

One suggested it was not natural but human resources that were lacking: "No man of property and enterprise has yet pitched his tent here so as to give the town and trade of the port a beginning, although it is certainly the most eligible situation upon the Island."[23] It may have been that the distance from Charlottetown or other communities on the Island was a discouragement. Eventually, however, Georgetown found its footing. An important nudge came in 1839 when it was designated a "free port" (which, as the name suggests, permitted shipping to British, American, and West Indian ports) and, domestically, given a subsidy for the beginning of packet and ferry service from Georgetown to Pictou, Nova Scotia. The pace of things accelerated quickly; within a few years, Georgetown was second only to Charlottetown as the Island's busiest port.

Fishing and farming remained the most prominent occupations in the region, but shipping brought important goods and capital to the storefronts of Kent and Water Streets. An 1864 directory shows a robust and relatively diverse economy in the town, listing white-collar professionals (accountant, surveyor, teacher), commercial businesses (hotel, grocery, drugstore), and skilled trades (ship repair, blacksmith, carpentry).[24] American fishing fleets filled the harbour, especially in the years of reciprocity with the United States.[25] Georgetown, with its neighbours of Cardigan and Montague, also became an important shipbuilding centre; indeed, the industry held on in the Three Rivers region through to the early twentieth century, later than the rest of the Island, despite the perennial uncertainties of global markets or local weather.[26] With the county institutions of a courthouse, market hall, and school, and prosperous streetscape architecture, Georgetown had gathered the trappings of a respectable Victorian town.

And yet, there was still some disappointment that Georgetown had not prospered *enough*. A visitor in 1861 described the enigma that the town posed to an era that celebrated improvement and enterprise: that a "harbor probably equal to any in the world, being capable of containing, in perfect safety, and by facility of approach, the entire navy of Great Britain" nevertheless sat as "the abode of apathy." It was a mystery "why the pleasantest associations cannot mingle there with the urbanities of life."[27] The enumerator for the 1871 census, on the other hand, was quite sure of the root of the problem. Neil Matheson lamented that "if [Georgetown] only had the water accommodations to which its natural advantages entitle it, it would not only give an impetus to the trade of the Town but benefit the Eastern part of the Island in general."[28] By "water accommodations," of course, he meant bigger and better port facilities – in other words, improvements.

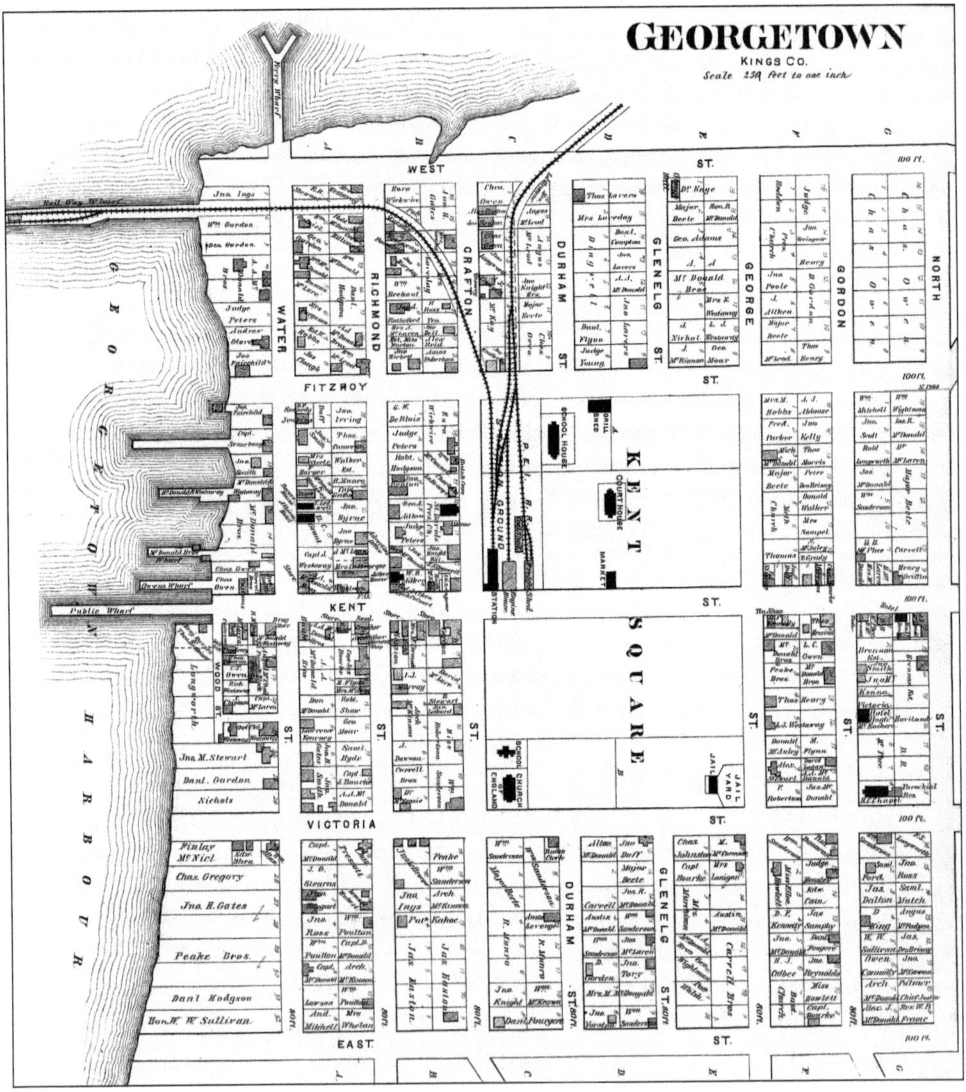

Georgetown, *Meacham's Illustrated Historical Atlas of Prince Edward Island* (1880). | Island Imagined Map Collection, Robertson Library, UPEI

Confederation with Canada, and the subsequent arrival of the PEI Railway in Georgetown, proved something of a mixed blessing. It was actually the legislative representative from Georgetown – Andrew Archibald Macdonald, who had layered mercantile wealth from the carrying trade and mackerel fishery on top of familial lands in the county – who attended the Confederation conferences and became the Island's only Father of Confederation, after it became clear that the PEI Railway was about to bankrupt the colony.[29] The railway did support a coordinated system of transportation, first with the ferries to Pictou and Halifax and then with the federal icebreakers across the Northumberland Strait. It also made Georgetown somewhat more accessible overland and viable as a secondary tourist destination. Now the quiet was a commodity; "somnolent Georgetown" could be presented as a soothing respite rather than a commercial failure.[30] In this, however, it was not particularly distinct from the rest of the Island and unable to compete with the more popular beachfront destinations along the north shore. A typical guidebook praised the "handsome and delightful capital Charlottetown" and added simply, "At the eastern end of the Island are Souris and Georgetown."[31] As the 1880 map on page 146 shows, the railroad was a significant intrusion, crashing into the grid and claiming the central square for its coal shed and engine house in ways that must not have been terribly soothing or somnolent.

Like many coastal communities, Georgetown found itself shunted to the margins when the highways of water were superseded by the highways of asphalt. Its out-of-the-way location and its remarkable harbour still held some advantages – notably for rumrunners eluding authorities during Prohibition in the United States – but its viability as a port remained a perennial question for the provincial government. When the Georgetown ferry was replaced by two other ferry routes to the mainland, it left "the fading seaport on its longer finger of land a dead-end destination."[32] It was also exactly what the alphabet soup of government agencies devoted to industrial development (notably the Department of Rural Economic Expansion) and Crown corporations (in this case, the Prince Edward Island Industrial Enterprises) had in mind as the kind of rural community that would be well-served by state intervention and "innovation."

Georgetown won enduring notoriety in Island history for the "debacle" (or, "corporate robbery") of an excruciating and expensive attempt in the 1960s to (re)establish a fish processing plant and shipyard on the waterfront. To be fair, these were reasonable propositions given the historical antecedents of nineteenth-century Georgetown. The Commission of Inquiry struck in 1969 to investigate the situation noted that these plans "were intended to provide industrial wealth to the Province, employment to its people, and to restore the town of Georgetown to its glory of the past century."[33] But the projects were so badly researched, planned, and managed, driven by external and speculative interests with no real connection to the town itself, they promptly went bankrupt, leaving the province entangled in the properties for decades and ultimately millions of dollars out of pocket. Today, none of the large-scale operations have survived and the Georgetown waterfront has become something of a cautionary tale in a region full of such projects.[34] After years of ownership cycling through some of Atlantic Canada's largest corporations (J.D. Irving, Cavendish Farms) and back to the province, it might seem, as historian Robert Tuck once suggested, that "the region, like Georgetown, has been betrayed by progress."[35] The town's population today is half what it was at the turn of the twentieth century, and significantly smaller than either Lunenburg or Saint Andrews.[36]

But we have short memories when it comes to failure and a resolute faith in development. Today, the amalgamated municipality of Three Rivers advertises its "tremendous potential for growth." Most recently, the province and the University of Prince Edward Island have sponsored the construction of a new industrial park, CleanTech Park, to host companies investing in renewable energies and carbon-reduction technology as well as some postsecondary programs related to clean technology. While the impulse is laudable, and the issue critical, the goals are familiar: to diversify the economy with "ingenuity and enterprise" and "create a new sector for growth on PEI."[37] But will people *live* there because of CleanTech? And is there any role for the waterfront?

Saint Andrews

Like Lunenburg and Georgetown, Saint Andrews was the product of eighteenth-century British concerns, but presented more clearly an immediate solution to an immediate problem. The townsite sits at the mouth of the St. Croix River, where a long peninsula juts into Passamaquoddy Bay, on the traditional territory of the Peskotomuhkati. The Peskotomuhkati (along with the Mi'kmaq and Maliseet across the Acadian peninsula) had agreed to various treaties of peace and friendship, or "peace and accommodation," with the British following the Seven Years War.[38] But it was another treaty that brought the site more directly into British view.

A few New Englanders had already occupied the shores of Passamaquoddy Bay, drawn by the rich lumber stands, but its proximity to Maine took on new importance in the fall of 1783. The Treaty of Paris identified the St. Croix River as part of the boundary between the new United States and British North America ("a line to be drawn along the middle of the River St. Croix, from its mouth in the Bay of the Fundy to its source," according to the treaty). This stretch of the border would be one of the most contentious aspects of British-American relations for another half-century, in part because of the value of the white pine that lay in the Maine and New Brunswick uplands and in part because of the difficult terrain. Finding the exact route of "the true St. Croix" – and thus the international border – would plague surveyors and diplomats for decades, but the presence of a Loyalist community at the entrance to Passamaquoddy Bay helped anchor the British claim of the St. Croix route and marked the edge of British America.[39]

Like Lunenburg, Saint Andrews was force-grown. Hundreds of Loyalists from the Penobscot region, along with several disbanded regiments, began arriving by the fall of 1783. The town was surveyed and allotted in 1784 in the signature Morris grid and with street names that announced allegiance to the royal family and British governors. Instead of a rural royalty, settlers at Saint Andrews and environs were awarded large tracts of forest at the head of Passamaquoddy Bay.[40] Residents quickly developed a sense of economic possibility based on the town's situation. As Morris's notations make clear, the stands of trees – including "beach [sic], birch, hemlock, spruce, and firr [sic]," or "beach [sic], birch, poplar, elm, pine, spruce, and maple" – were immediately within view, as were the "fine natural mill seats" on the rivers; indeed, Morris repeats several times the special value of the pine. In May 1784, barely six months after the arrival of the first transports, a group of petitioners wrote:

> In the Grand Bay of Passamaquoddy alone a sufficient Quantity of Board and other Lumber can in a short time be furnished to supply the greatest part of the British West India Islands: likewise large quantitys [sic] of masts, spars & other Lumber suitable for the European Market. Altho' the first of our Settlers only arrived here in October last, yet we have already sent a number of Cargoes of Lumber to the West Indies and several ports in Nova Scotia, and as more Saw Mills are now erecting our Exports of Lumber will rapidly increase.[41]

The exports of lumber did increase, along with masts and spars and ships – all of which proved useful in what soon became Saint Andrews'

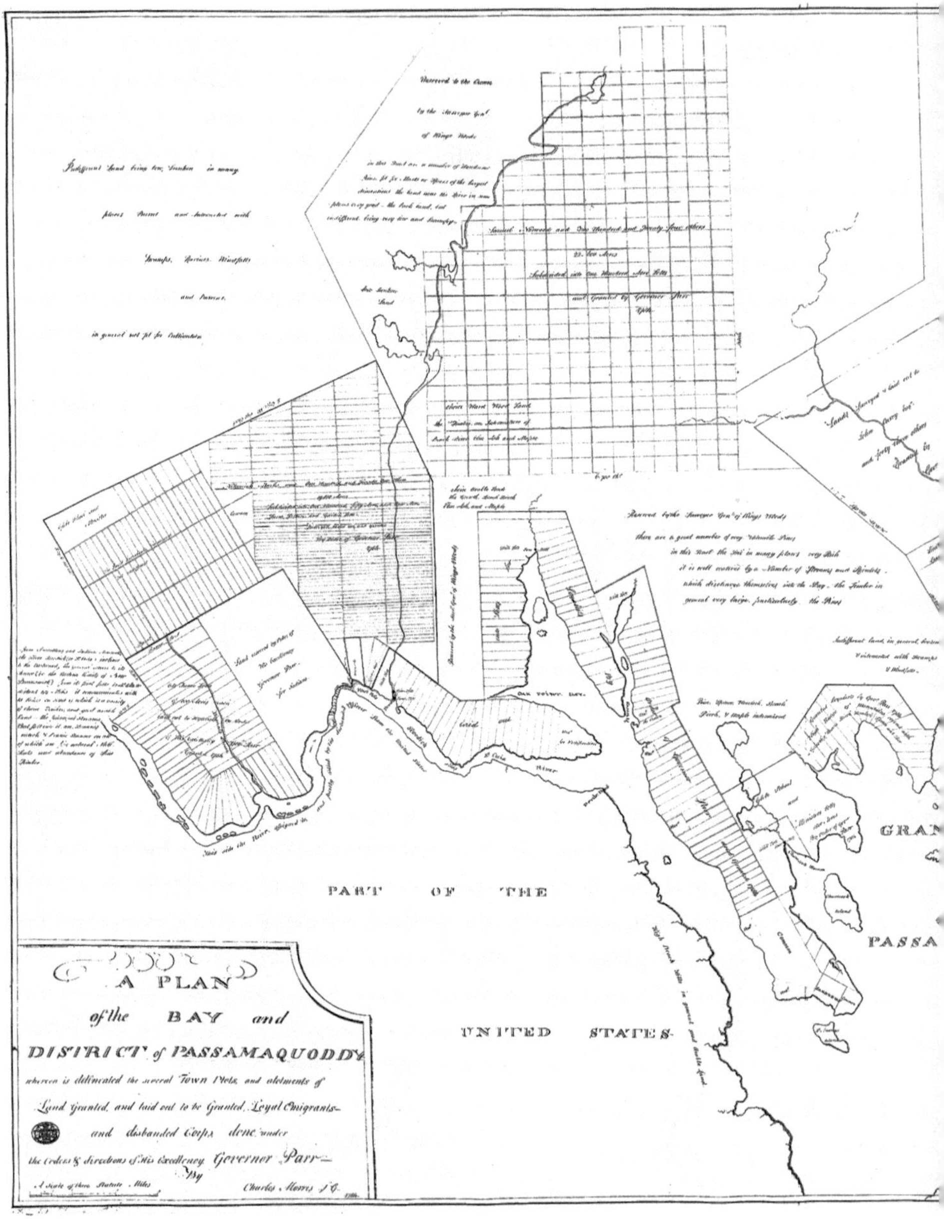

"A plan of the Bay and District of Passamaquoddy whereon is delineated the several town plots, and alotments [sic] of land granted, and laid out to be granted, loyal emigrants and disbanded corps done under the orders & directions of His Excellency, Governor Parr by Charles Morris, S.G. 1784" (detail). | British Library maps_k_top_119_51

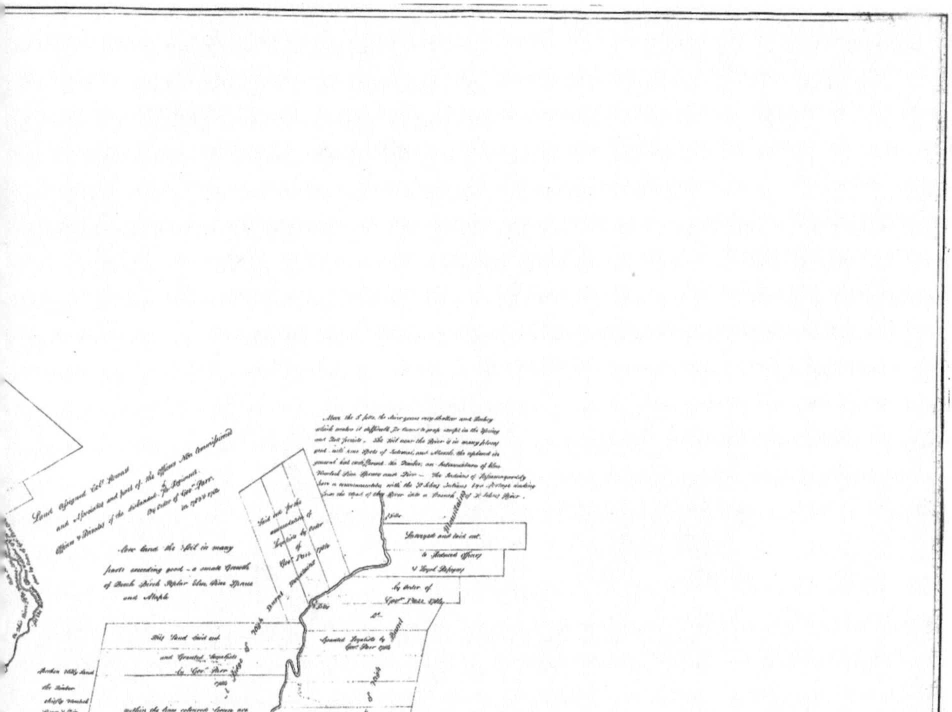

AY
ODDY
BAY OF FUNDY
Great Lake, hereabouts not yet Surveyed.

other economic specialty. As a customs port, Saint Andrews could welcome British and American ships in peace and quietly liaise between the two economies in war, navigating between British protections, American markets, and international waters to profit from "transshipping" (or, depending on the legality, smuggling).[42] Loyalist and American towns faced each other across the water and along the border (St. Stephen/Calais, Saint Andrews/Eastport). There would be periods of anxiety – as during the War of 1812, when blockhouses were built on the edges of the town, or the Fenian scare of 1866 – but American proximity would largely be forgiven and forgotten with Saint Andrews' "renaissance" as a tourist destination. Like Georgetown, Saint Andrews would find itself fading as an international port in the age of rail; it was bypassed as the Atlantic port of the St. Lawrence first in favour of Portland and then by Saint John. But the piecemeal construction of spur lines in New Brunswick eventually led into the town, and then the tide began to turn.[43]

By the 1860s, the well-to-do from American cities had begun establishing summer homes and camps along the islands and coastlines of Maine, including the Passamaquoddy island of Campobello. At the same time, some prominent Canadian figures – including a few Fathers of Confederation – had begun taking summer trips to Saint Andrews. Across eastern North America, other rural and resource regions were being remade in similar ways as cottage country or recreational hinterlands.[44] While there were half-hearted attempts to repackage Georgetown's lassitude as touristic repose, the "sleeping beauty" of Saint Andrews was awakened by the kiss of railway and steamer links to the monied centres of Montreal and Boston and the rather self-congratulatory presence of "American enterprise" or "Americans and railroad men."[45] In 1888, a group calling itself the St. Andrews Land Company (what local residents called "The American Syndicate") bought parcels of land through and around the town plat and at Indian Point. Initially the company promised to build summer homes "in the choicest locations ... spots from every one of which panoramic views are had," as well as a water system for the town and extensive landscaping for a park of riding paths and promenades.[46] The angle of the map on page 153 presents the peninsula tucked below the uplands, suggesting both a destination (the train running along the outer edge) and an elite preserve. Ultimately the syndicate remained content to simply sell the lots, leaving the construction of summer homes to private families and architects, but it did open the luxurious Algonquin Hotel as well as a club headquarters on Water Street and a golf course on the western edge of town.[47]

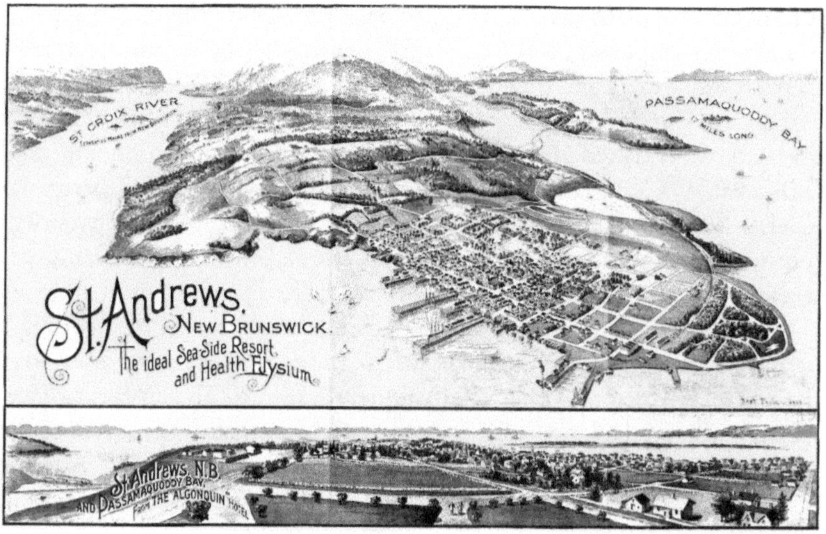

Bert Poole, *St. Andrews, New Brunswick. The ideal Sea Side Resort and Health Elysium* (ca. 1890). | Tides Institute and Museum of Art

The second major infusion of "enterprise" came shortly after the Algonquin opened, when the Canadian Pacific Railway (CPR) acquired the right-of-way on the line, built a major transfer station at McAdam, and began delivering capital, publicity, and clientele to Saint Andrews. This included company president Sir William Van Horne, who purchased Ministers Island on the far side of the peninsula as a private estate. The CPR consolidated the image of the town as a destination by purchasing the Algonquin in 1903 and rebranding the town as St. Andrews-by-the-Sea.[48] Though tourist brochures insisted that the town welcomed guests beyond purely "society," the building of new and elaborate summer homes, the piazzas of the Algonquin poised on the crest of the hill, and the frequent newspaper accounts of prominent figures staying at the hotel or in town quickly established a particularly highbrow image for what had now become a resort town. As the CPR described it, "The little village which once had great aspirations to commercial importance, but is now happily content with the prestige and supremacy it has attained as a charming summering-place."[49]

Luxury and convenience had successfully made quiet a marketable commodity; "its quiet and restfulness, together with the embellishment which Nature and man have given it, constituting its principal charm for summer-visitors." In the rechristened St. Andrews-by-the-Sea, there was a consistent emphasis on the healthful qualities of the location, whether the restorative calm and quiet (far from "the cares of business and the heat and dust of the city"), regular drainage of sewage with the tides, the "pebbly walks and drives" through pine and sea air, or the moderate and comfortable "salubrity of climate." The sea breezes promised particular relief as "an elysium for the hay fever patient."[50] By the turn of the twentieth century, the economy of Saint Andrews seemed firmly that of summer service.

A seasonal (if perennial) infusion of capital and taste is not, by itself, a path to economic equilibrium or sustainability, but Saint Andrews had the advantage of three other institutional moorings. The customs office, which had played a larger role in the nineteenth century, remained at the waterfront. In 1908, the Dominion government established the Atlantic Biological Station, Canada's first permanent marine research location. While the rich ecologies of the Fundy zone were both scientifically interesting and economically important, historian Eric Mills has pointed out that the "social factors, and ease of access from central Canada" – that is, a direct railroad link – were key to the decision to locate the station at Saint Andrews.[51] In 1969, a consortium of Canadian universities and federal departments opened the Huntsman Marine Science Centre, as an expanded teaching and research facility, on the grounds of an Edwardian summer estate built for one of "the railroad men." Finally, amid Canada's centennial enthusiasm for historic reconstruction, the National Historic Parks and Sites Branch acquired and restored the surviving blockhouse: a small signifier of Loyalist, or gently not-American, sentiment that would play well in the national narrative. Later, Saint Andrews was one of several small towns in the Maritimes swept up by the wave of site designation for national historic districts that encompassed eighteenth-century town grids.[52] Together, the combined federal and academic interests brought some much-needed diversity of interest and audience to the town.

Accordingly, of these three communities Saint Andrews has arguably weathered the postindustrial turn with the greatest economic and demographic equanimity. While the town's population dipped in the early twentieth century, it has recovered to late-Victorian levels; in 2021 there were about as many residents as in 1881. There is still a seasonal population, but it is a minority; in addition, Saint Andrews has avoided the crushing numbers of daytrip visitors swinging through Lunenburg from Halifax.[53] Although not immune to the era of bureaucratic intervention in the cause of rural economic development, the town has avoided dramatic experiments (or failures) in imported industrial ventures. (The provincial government had to rescue the Algonquin financially in 1973 and bought the property in 1984, though it was still marketed as "Canadian Pacific Hotels and Resorts." The province then sold the hotel in 2012 to the American Marriott Hotel chain, in a sense returning the property to an "American syndicate." The town has also sidestepped large-scale infrastructure projects, despite occasional proposals for tidal generation or natural gas terminals. Indeed, the lasting presence of the Algonquin has elevated infrastructure into something quaint and nostalgic, as in the aesthetic of Edwardian railway design at the McAdam Junction station. Saint Andrews markets much the same attractions – fishing, yachting, and golf – as it did over a century ago and regularly appears in lists of "best towns to visit" or "best summer destinations" in Canada. Staying small has had value.

The town is vulnerable, though, in the new climate era. There is significant concern about rising tides and flooding on the low-lying peninsula.[54] Despite some acknowledgment of the value of green space throughout the town, as "natural infrastructure" to absorb higher water, it is here, somewhat ironically, that the small town sounds more like its city cousins. In 2024 the town opted for infill to buttress Market Wharf and to explore hard engineering with a sea wall.[55] This may offer some protection in the short term, but by this point in our story, we might wonder if another engineering intervention is really the solution.

We can't all live in Saint Andrews. And not every place can, or should, *be* like Saint Andrews. But (despite the emphasis of this book) towns like these three are more numerous in and more characteristic of Atlantic Canada than its large port cities; think of Maitland, Annapolis Royal, Shelburne, or Guysborough just in mainland Nova Scotia alone. Can we reorient our thinking – our preoccupations of the industrial era and the urban turn – to recentre these kinds of communities? Why do they thrive where they do? When is a limit to growth optimal and sustainable? While this book has been primarily concerned with the force and costs of some traditions – the learned habits and practices of waterfront development that are reinscribed in our cities over decades – history also suggests other possibilities, paths that might help us imagine alternatives for how else we might live beside a changing coast. We can look at places like Saint Andrews, Georgetown, and Lunenburg to identify the qualities that make communities appealing, viable, and distinctive – qualities that sustain environmental integrity and our well-being – qualities that we can protect in small places and perhaps reintroduce in bigger ones, especially as we try to anticipate and adapt to climate change. The past encompasses these different moments; what we need to do now is stretch our imagination forward to different possibilities.[56]

One factor would be the human scale along the streetscapes and within the community as a whole. This includes a scale of movement that does not rely on cars and spaces that invite human movement in public and green spaces so people feel an attachment to the town *as* a space; what the Lunenburg report of 1966 called "a scale of contact with the sea and the land behind." This includes, crucially, views of the sea: the ability to stand at the water's edge, without the barricade of a chain-link fence or the impersonal distance of a concrete harbour apron. We have seen at historic sites – from the Saint Andrews blockhouse to the Halifax Citadel – the value of view planes, the ability to situate ourselves and historic pathways in the harbour geography. Being aware of the coast, its water and weather, is not just aesthetically compelling but reminds us of the environmental and historical circumstances that made these places possible, their historical logics, and the inescapable effects of climate change actively reshaping them in our lifetimes.

Human scale means equilibrium rather than an ethos of growth. While nineteenth-century boosters lamented the lack of enterprise, today that sensibility of calm is highly attractive. While we are seeing important experiments in density in city neighbourhoods – to supply much-needed housing without suburban dependencies on petroleum for transportation – it is important to also remember the physical benefits of what Parks Canada identifies at Saint Andrews as "the generous proportion of landscape to built elements."[57] As New Brunswick naturalist David Christie wrote, "City people should not be divorced from nature. They need places to walk through the trees, to sit and watch the ocean waves."[58] These three towns have all retained their wide streets, preserving what Birch Bagster described at Georgetown as "a spaciousness well calculated to gratify and surprise the visitor."[59] None of these places have a gas-station strip, and all are the better for it.

Of course, viable, multigenerational communities need to allow for contemporary lives within the period streetscape. As historian Ed MacDonald (himself from Newport, PEI, across the Cardigan River from Georgetown) observed, if you build things for the people who live there, people will want to live there.[60] A student from Annapolis Royal once told me he wanted to move back home after graduation, but unless he worked as an historical interpreter or "in a pottery shop" he didn't see many options. In the post-industrial era many of these small communities have struggled to establish a means of living between the wharf and the gift shop, between the extractive economy and the service one. Both, in a way, have meant selling the place itself: whether the product is shipped out or the market is brought in. The knot has been finding sources, plural, of sustainable income: income that does not rely on harvesting down a limited resource or relying on external capital. It means weaving a richer and more diverse economic landscape than a manufacturing plant or a hotel. This is not a recent revelation; even at the turn of the twentieth century a commentator called on residents of Lunenburg to "encourage home industries" in town, or no further than Halifax, rather than sending their money to Toronto.[61] The most successful or sustainable projects seem to find value in *learning* from the place rather than consuming it; whether historic narratives, architectural design, or scientific research. History suggests that projects planted "from away" rarely take root, either economically or in community affection. But the COVID-19 pandemic demonstrated that we are also capable of working from away in ways that situate productivity in healthier and more balanced lives. These communities may showcase how it is possible to choose first where to live, and then, how to work.

Crucially, this means finding space to move and respond as the water returns in this new climate era. This does ask us to accept physical or topographical constraints: accepting where the water is, anticipating where it may be, rather than assuming we can control its reach or add land where we want. More challenging, it means accepting limits on ourselves, what we need, want, and require from our daily environments – what, exactly, is important and necessary in what Bagster, visiting Georgetown in 1861, called "the urbanities of life." As Graeme Wynn has observed, we need to acknowledge the environmental costs of our choices and recalibrate our expectations.[62] Rather than always building out, can we value what we have? It is often difficult to convince people that such a course is nothing more than discomfort or denial: sacrificing growth and convenience appears to be a disavowal of technological progress and the achievements of modernity. But it is more than that. It means counting environmental well-being – quality of life for ourselves, others, and the natural world – as an essential *part* of progress. I love a good episode of *Road to Avonlea* as much as anyone, but I wouldn't want to live in a rural community without a hospital, library, or high-speed internet. At the same time, Avonlea did have many of the things we materially need – a school, a hardware store, and the Island newspaper – set in a beautiful place. The goal is a balance between self-reliance and connection, between locally supplied goods and services and the benefits of a wider geography. In many ways, that *was* the nineteenth-century coastal community. It may allow us to live by the sea.

Can finding the foreshore again help us imagine healthier, more resilient cities? The "creative city," the "fluid city," the "porous city": these and other framings suggest a living, adaptable urban community with blue and green spaces throughout. The challenge, as Isaac Land noted, will be "to preserve the elements that make cities livable and functional in a state of partial inundation."[63] We should certainly strive to reconnect people to older water and landscapes wherever possible, to supply places of congregation and contemplation. But plans for "waterfront regeneration" need to acknowledge the inheritance of history amid such promises of "transformation." For one, history has set the terms, like "development," that we still traffic in and the tools we have to work with. It also requires us to consider *why* our predecessors acted as they did, to acknowledge that we have likely benefited from the quest for "improvement," and to ask: Was it worth it? At the same time, we find sympathetic figures (like George Robinson or W.C. Harris) who remind us that these are not new or inevitable decisions. As coastal communities of *all* sizes struggle with questions of sustainability, we might look to the urban past for points of comparison, connection, and alternatives.[64]

Standing at the end of railroad steel in the lower cove of Saint John, looking to the pinpoint light at Partridge Island. Climbing down the steps to the old King's Beach in St. John's, with the silhouette of Cabot Tower in the distance. Finding the unexpected marsh at the foot of the Confederation Trail in Charlottetown. Hearing fog horns through silver mist on the rocky beach at Point Pleasant Park at the edge of the Halifax peninsula. These are places and moments only possible by the sea, places and moments that help us understand our environmental history. In its aesthetic, the waterfront reflects the relationship with nature we have chosen. In its infrastructure, it makes visible the cost of nation-building. In its location, it is the part of our cities affected by climate change in dramatic and daily ways. In its complexity, as a meeting of different kinds of nature, it reminds us that we are part of a larger whole. All of these – past, present, future – ask us to use our imagination.

Notes

Foreword: On the Waterfront

1 Tracey Wade, *Overview of Canadian Communities Exposed to Sea Level Rise* (National Collaborating Centre for Environmental Health, 2022), https://ncceh.ca/resources/evidence-reviews/overview-canadian-communities-exposed-sea-level-rise.

2 William Cowper, *The Task: A Poem, in Six Books,* Book 1, *The SOFA* (Printed for J. Johnson, 1785), line 749.

3 George Perkins Marsh, *Man and Nature: Or, Physical Geography as Modified by Human Action* (Charles Scribner and Co., 1864), iii.

4 A notable example is the Wenner-Gren–sponsored invitational symposium of 1955, with seventy participants selected "for their common interest and curiosity about what man has been doing to and with his habitat." This produced a 1,232-page book: William L. Thomas Jr., ed., with the collaboration of Carl O. Sauer, Marston Bates, and Lewis Mumford, *Man's Role in Changing the Face of the Earth* (University of Chicago Press, 1956). See also Donald Worster, ed., *The Ends of the Earth: Perspectives on Modern Environmental History* (Cambridge University Press, 1989), and John R. MacNeill, "Observations on the Nature and Culture of Environmental History," *History and Theory* 42, 4 (2003): 5–43.

5 Matthew Klingle, *Emerald City: An Environmental History of Seattle* (Yale University Press, 2007); Matthew Gandy, *Concrete and Clay: Reworking Nature in New York City* (MIT Press, 2003); and at a different scale and in different registers, T.R. Oke, "City Size and the Urban Heat Island," *Atmospheric Environment* 7, 8 (1973): 769–79; Richard T.T. Forman, *Urban Ecology: Science of Cities* (Cambridge University Press, 2014); and Matthew Gandy, *Natura Urbana: Ecological Constellations in Urban Space* (MIT Press, 2022).

6 Mark Williams, "Out of the Deep: 10 Massive Land Reclamation Projects," *Modus,* 6 February 2024, https://ww3.rics.org/uk/en/modus/natural-environment/land/out-of-the-deep--7-massive-land-reclamation-projects--.html.

7 There are, probably, only five countries with more than 500 square kilometres of reclaimed land. Although reclaimed land accounts, famously, for about 20 percent of *terra firma* in

the Netherlands, its 7,000 square kilometres place it second to China, which has twice that area. South Korea counts more than 1,500, the United States just tops the four-figure mark, and Japan stands fifth, with slightly more than 500 square kilometres. It may soon be surpassed by the United Arab Emirates, which has 470 square kilometres of reclaimed land.

8 Robert J. Hoeksema, "Three Stages in the History of Land Reclamation in the Netherlands," *Irrigation and Drainage* 56, S1 (2007): S113–26.

9 "Fisherman's Wharf," Stan Rogers, SOCAN 1976. Used by permission.

10 S. Arnold and A. Fenech, *Prince Edward Island Climate Change Adaptation Recommendations Report* (University of Prince Edward Island Climate Lab, 2017), https://projects.upei.ca/climate/files/2018/10/PEI-Climate-Change-Adaptation-Recommendations-Report_medres.pdf.

11 This sophisticated model "combines historical erosion data, IPCC [Intergovernmental Panel on Climate Change] model projections of future sea-level rise, aerial imagery, and high-resolution digital elevation data to develop analytical visualizations of coastal erosion and future sea-level-rise scenarios." A brief account of CLIVE and a video clip showcasing it are available at https://projects.upei.ca/climate/clive/.

12 Professor Stephen Sheppard of the Centre for Advanced Landscape Planning at UBC has been involved with most of these collaborative projects. His contributions are conveniently represented in Stephen Sheppard, *Visualizing Climate Change: A Guide to Visual Communication of Climate Change and Developing Local Solutions* (Earthscan, 2011).

13 Details of Future Delta 2.0 and a downloadable version of the game are available at https://www.futuredelta2.ca/.

14 Sheppard, *Visualizing Climate Change;* Aleksandra Dulic, Jeannette Angel, and Stephen Sheppard, "Designing Futures: Inquiry in Climate Change Communication," *Futures* 81 (February 2016): 1–23.

15 Dulic, Angel, and Sheppard, "Designing Futures," includes discussion of the challenges of science communication and the value of imagined, virtual, and poetic "spaces" in facilitating such communication, and provides a useful summary of the design and structure of the Future Delta game.

16 Nancy Russell, "Climate Change Simulator Tool Draws Gasps, Even Tears from P.E.I. Residents," *CBC News,* 4 July 2024, https://www.cbc.ca/news/canada/prince-edward-island/pei-clive-climate-change-simulator-updated-1.7253461.

17 Robert Vaughan, *The Age Of Great Cities: Or Modern Civilization Viewed In Its Relation To Intelligence, Morals And Religion* (Jackson and Walford, 1843).

18 Examples abound. The WUI example is drawn from discussion of the BC government's "Wildland Urban Interface Risk Class Maps" (https://www2.gov.bc.ca/gov/content/safety/wildfire-status/prevention/fire-fuel-management/wui-risk-class-maps). The Municipal Climate Services Collaborative (MCSC) developed a series of statements about climate change resilience, including one pertaining to Saint John: *Using Climate Information to Drive Adaptation: Local Government Case Studies from Across Canada* (October 2021), https://changingclimate.ca/site/assets/uploads/2022/01/MCSC-Climate-Change-Resilience-Case-Studies-presentation-Oct-15-21-ENG.pdf. Natural Resources Canada, Climate Change Impacts and Adaptation Program, *Adapting to Sea Level Rise in Charlottetown, P.E.I.* (2013), https://publications.gc.ca/collections/collection_2014/rncan-nrcan/M174-10-2013-eng.pdf.

19 Dulic, Angel, and Sheppard, "Designing Futures," 22, 15, 8, 16–19.

20 For a thought-provoking commentary on the "stream of tendency" notion, see Leslie Kendrick, "Literary Influence on Legal Philosophy: A Comment on Professor Meyler's *Cardozo's Literary Precedents*," *Yale Journal of Law and the Humanities* 34, 1 (2023): 246–52.

21 Northrop Frye, *The Educated Imagination* (Indiana University Press, 1964), 23, distinguished between (intellectual) Science and (emotional) Art: "Science begins with the world we have to live in, accepting its data and trying to explain its laws ... Art, on the other hand, begins with the world we construct, not with the world we see. It starts with the imagination, and then works toward ordinary experience." Few hold as tightly to this distinction today, but its emphasis on the importance of the imagination's "power of constructing possible models of human experience" is salient.

22 The "islands of sanity" idea seems to be enjoying a "moment" after the tumultuous months of 2025. This might be attributable, in part, to Margaret J. Wheatley, *Restoring Sanity: Practices to Awaken Generosity, Creativity and Kindness in Ourselves and Our Organizations* (Berrett-Koehler, 2024). See also https://margaretwheatley.com/wp-content/uploads/2024/08/Wheatley-School-Administratorpdf.pdf; and for the "resilience" quote, Till Leinen, "Creating an Island of Sanity in This Crazy World," *Resilience*, 13 November 2024, https://www.resilience.org/stories/2024-11-13/creating-an-island-of-sanity-in-this-crazy-world/.

23 William Wordsworth, Untitled Sonnet 18 ("The world is too much with us"), *Poems in Two Volumes*, vol. 1 (Longman, Hurst, Rees, and Orme, 1807), 122.

24 See also Graeme Wynn, "The Power of Words and the Tides of History: Reflections on *Man and Nature* and *Silent Spring*," in *Geographies of Knowledge and Power*, ed. Peter Meusberger, D. Gregory, and L. Suarsana (Springer, 2015).

25 A quarter-century ago, Jill Ker Conway, Kenneth Kenniston, and Leo Marx, eds., *Earth, Air, Fire, Water* (University of Massachusetts Press, 1999), lamented the de facto marginalization of the humanities in environmental scholarship and sought to remedy this. See also Poul Holm and Verena Winiwarter, "Climate Change Studies and the Human Sciences," *Global and Planetary Change* 156 (2017): 115–22; John Tosh, *The Pursuit of History: Aims, Methods, and New Directions in the Study of Modern History*, 4th ed. (Longman, 2006), 28ff.

Acknowledgments

1 L.M. Montgomery, *Jane of Lantern Hill* (Seal Books/Random House of Canada, 1993), 147.

Introduction: The Vanished Shoreline

1 Andrew Lipman, *The Saltwater Frontier: Indians and the Contest for the American Coast* (Yale University Press, 2015), x.

2 Jared Orsi, *Hazardous Metropolis: Flooding and Urban Ecology in Los Angeles* (University of California Press, 2004).

3 Ted Steinberg, "Afterword," in *Environmental Disaster in the Gulf South: Two Centuries of Catastrophe, Risk, and Resilience*, ed. Cindy Ermus (LSU Press, 2018), 184; and Jeff Goodell, *The Water Will Come: Rising Seas, Sinking Cities, and the Remaking of the Civilized World* (Little, Brown, 2017), 146.

4 "Halifax Is Sinking," *Dalhousie Gazette*, 20 November 2010.

5 The term "littoral," though less common a term to describe shore or coastline in English, is actually "coastline" in French.

6 I mean burdensome literally; in 2023, scientists reported that New York City, built on an estuarine archipelago, is sinking under the weight of its buildings. Tom Parsons, Pei-Chin Wu, Meng (Matt) Wei, and Steven D'Hondt, "The Weight of New York City: Possible Contributions to Subsidence from Anthropogenic Sources," *Earth's Future* 11 (2023): 1–13.

7 Graeme Wynn, "Reflections on the Environmental History of Atlantic Canada," in *Land and Sea: Environmental History in Atlantic Canada,* ed. Claire Campbell and Robert Summerby-Murray (Acadiensis Press, 2013), 243ff.

8 Donald Worster, "Ice, Worms, and Dirt: The Power of Nature in North American History," in *Method and Meaning in Canadian Environmental History,* ed. Alan MacEachern and William J. Turkel (Nelson, 2009), 31.

9 Lauren MacDowell, *An Environmental History of Canada* (UBC Press, 2016), 6.

10 George Perkins Marsh, *Man and Nature: Or, Physical Geography as Modified by Human Action* (S. Low, Son and Marston, 1864), 13, 330–32.

11 R.L. Sherlock, *Man as a Geological Agent* (H.F. & G. Witherby, 1922), 238–58; and "The Influence of Man as an Agent in Geographical Change," *The Geographical Journal* 61, 4 (1923): 258–68.

12 Paul Crutzen is credited with coining the term – and dating it to the late eighteenth century (and the adoption of fossil fuels on an industrial scale) – in a piece titled "Have We Entered the Anthropocene?," *International Geosphere-Biosphere Programme* (2000). Others, such as J.R. McNeill, have argued for a "great acceleration" from the mid-twentieth century. McNeill also offers a diplomatic compromise that reconciles these – and that I follow – in seeing the Anthropocene as an epoch with stages of scale and acceleration within it: "It may have begun about 1800, but the scale, scope, and pace of anthropogenic environmental change all grew enormously from the mid-twentieth century." J.R. McNeill, "Introductory Remarks: The Anthropocene and the Eighteenth Century," *Eighteenth-Century Studies* 49, 2 (2016): 117–28; and with Peter Engelke, *The Great Acceleration: An Environmental History of the Anthropocene Since 1945* (Harvard University Press, 2016).

Scholars continue to debate the meaning of the Anthropocene, named for the "age of humans": when it began, and by extension or more precisely, the human action (and impact) that defines it. Was it when humans began dedicated agriculture and altered the soil? Or when we detonated an atomic device and altered the atmosphere? Was it the industrial revolutions beginning in the late eighteenth century, or their global acceleration in the mid-twentieth? Was it the turn to fossil fuels? Geologists and stratigraphers remain divided over the evidence of a geological strata that registers human presence, while environmental humanists have crafted different metaphors to highlight the primary forces within the Anthropocene, be they capitalism, imperialism, or technology. On the multiple narratives of the concept, see McNeill, above; Christophe Bonneuil, "The Geological Turn: Narratives of the Anthropocene," in *The Anthropocene and the Global Environmental Crisis,* ed. Clive Hamilton, Françoise Gemenne, and Christophe Bonneuil (Routledge, 2015); and Jeremy Davies, *The Birth of the Anthropocene* (University of California Press, 2016). From the geologists, see the Subcommission on Quaternary Stratigraphy; and from the geographers, "The Anthropocene," ed. David R. Butler, special issue, *Annals of the American Association of Geographers* 111, 3 (2021), especially Jeffery Hoelle and Nicholas C. Kawa, "Placing the Anthropos in Anthropocene," 655–62. As examples of the humanist reflection, see Donna Haraway, "Anthropocene, Capitalocene, Plantationocene, Chthulucene: Making Kin," *Environmental Humanities* 6, 1 (2015): 159–65; Steve Mentz, *Break Up the Anthropocene* (University of Minnesota Press, 2019); and Jeremy Chow, "Introduction: Eighteenth Century

+ Environmental Humanities," *Eighteenth-Century Humanities* (Rutgers/Bucknell University Press, 2023). I thank Maria Antonaccio, Ellen Herman, and Matthew Slater, my colleagues and co-instructors in the course on The Anthropocene at Bucknell University, for our discussions about the features and meanings of this topic.

13 A good introduction to some of these issues, from authors in Atlantic Canada, is an article by Ryan Guild, Xiuquan Wang, and Pedro A Quijón, all of the University of Prince Edward Island: "Climate Change Impacts on Coastal Ecosystems," *Environmental Research: Climate* 3, 4 (2024): 1–27.

Measures of sea-level rise vary, but see R. Daigle, "Sea-Level Rise and Flooding Estimates for New Brunswick Coastal Sections" (R.J. Daigle Enviro and the New Brunswick Department of Environment and Local Government, 2020); Bailey Brogan, Jamylynn McDonald, Blaine Lyons, Peter Johnston, and Graeme Stewart-Robertson, *Climate Change Adaptation Plan for Saint John* (Atlantic Coastal Action Program, 2020); T.S. James, C.M.I. Robin, J.A. Henton, and M.R. Craymer, "Relative Sea-Level Projections for Canada Based on the IPCC Fifth Assessment Report and the NAD83v70VG National Crustal Velocity Model," Geological Survey of Canada, Open File, 8764 (2021); Tracey Wade and CLIMAtlantic, "Sea Level Rise and Public Health Implications," National Collaborating Centre for Environmental Health (2022); Nova Scotia Department of Environment and Climate Change, *Halifax Census Division: Climate Risk Summary,* December 2022; and Adam Fenech, Xander Wang, Don Jardine, Ross Dwyer, Andrew MacDonald, Luke Meloche, and Catherine Kennedy, "Keeping Watch over Prince Edward Island's Coastline," *American Review of Canadian Studies* 55, 1 (2025). We are seeing more online mapping tools that allow us to visualize sea-level rise and other effects of climate change, such as climatedata.ca, chris.peiclimate.ca, and coastal.climatecentral.org.

14 J.B. MacKinnon, *The Once and Future World: Nature as It Was, as It Is, as It Could Be* (Houghton Mifflin Harcourt, 2013), 131.

15 Humanities scholars have, for at least a decade, highlighted the need to recognize that these questions are fundamentally ones of human motivation and action. See, for example, Poul Holm and Verena Winiwarter, "Climate Change Studies and the Human Sciences," *Global and Planetary Change* 156 (2017): 115–22; George C.D. Adamson, Matthew J. Hannaford, and Eleonora J. Rohland, "Re-thinking the Present: The Role of a Historical Focus in Climate Change Adaptation Research," *Global Environmental Change* 48 (2018): 195–205; and work by Julia Adney Thomas such as the forthcoming *The Historian's Task in the Anthropocene.*

16 Stephen Mosley, "Coastal Cities and Environmental Change," *Environment & History* 20, 4 (2014): 520.

17 David Harvey, *The Limits to Capital* (Verso, 2006), 378–79 and 428–29; Erica Schoenberger, "The Spatial Fix Revisited," *Antipode* 36, 3 (2004): 428–29; Carl Smith, *City Water, City Life: Water and the Infrastructure of Ideas in Urbanizing Philadelphia, Boston, and Chicago* (University of Chicago Press, 2014), 3–6; and Eric Sheppard, "The Spatiality of the Limits to Capital," *Antipode* 36, 3 (2004): 474.

18 Ashley Dawson, *Extreme Cities: The Peril and Promise of Urban Life in the Age of Climate Change* (Verso, 2017); see, for example, the works on New York below, and Bill Luckin and Peter Thorsheim, eds., *A Mighty Capital Under Threat: The Environmental History of London, 1800–2000* (University of Pittsburgh Press, 2020).

19 Matthew M. Booker, *Down by the Bay: San Francisco's History Between the Tides* (University of California Press, 2013), 1–5.

20 Verena Winiwarter, Gertrud Haidvogl, Severin Hohensinner, and Friedrich Hauer, "The Long-Term Evolution of Urban Waters and Their Nineteenth Century Transformation in European Cities. A Comparative Environmental History," *Water History* 8, 3 (2016): 209–33; M. Jasper Rubin, *A Negotiated Landscape: The Transformation of San Francisco's Waterfront Since 1950* (University of Pittsburgh Press, 2016), 16.

21 Rubin, *A Negotiated Landscape*, 3–5.

22 Matthew Gandy, *The Fabric of Space: Water, Modernity, and the Urban Imagination* (MIT Press, 2014), 1–2. Key works in American urban water history that have influenced my thinking include Joel Tarr, ed., *Devastation and Renewal: An Environmental History of Pittsburgh and Its Region* (University of Pittsburgh Press, 2003); Sherry Olson, whose metaphor of the "urban funnel" reads intake and disposal differently, in "Downwind, Downstream, Downtown: The Environmental Legacy in Baltimore and Montreal," *Environmental History* 12, 4 (2007): 845–66; Martin Melosi, *The Sanitary City: Environmental Services in Urban America from Colonial Times to the Present* (University of Pittsburgh Press, 2008); Steven Moga, *Urban Lowlands: A History of Neighborhoods, Poverty, and Planning* (University of Chicago Press, 2020); and Kathleen A. Brosnan, Ann Durkin Keating, and William C. Barnett, eds., *City of Lake and Prairie: Chicago's Environmental History* (University of Pittsburgh Press, 2020). Christina Dunbar-Hester, however, challenges the metabolism metaphor for its capacity to "naturalize" city infrastructure that is often harmful to biological life, in *Oil Beach: How Toxic Infrastructure Threatens Life in the Ports of Los Angeles and Beyond* (University of Chicago Press, 2023). From Canada, I would note Jennifer Bonnell's *Reclaiming the Don: An Environmental History of Toronto's Don River Valley* (University of Toronto Press, 2014) and Michèle Dagenais's *Montreal, City of Water: An Environmental History*, trans. Peter Feldstein (UBC Press, 2017). Edited collections have allowed for some comparative views, especially across international borders, as in Thaïsa Way, ed., *River Cities: City Rivers* (Dumbarton Oaks Research Library and Collection, 2018).

23 Traditionally, scholars have focused on the social and economic dynamics of port towns: for example, a special issue of *Early American Studies* 15, 4 (2017), and Brad Beaven, Karl Bell, and Robert James, eds., *Port Towns and Urban Cultures: International Histories of the Waterfront, 1700–2000* (Palgrave Macmillan, 2016). But the emergence of environmental history is evident in work from the American Atlantic seaboard; some examples that have influenced my thinking include Nancy Seasholes, *Gaining Ground: A History of Landmaking in Boston* (MIT Press, 2003); Michael Rawson, *Eden on the Charles: The Making of Boston* (Harvard University Press, 2010); Brian Black and Michael J. Chiarappa, eds., *Nature's Entrepôt: Philadelphia's Urban Sphere and Its Environmental Thresholds* (University of Pittsburgh Press, 2012); Mary Ryan, *Taking the Land to Make the City: A Bicoastal History of North America* (University of Texas Press, 2019); Kara Schlichting, *New York Recentered: Building the Metropolis from the Shore* (University of Chicago Press, 2019); and Pavla Simková, *Urban Archipelago: An Environmental History of the Boston Harbor Islands* (University of Massachusetts Press, 2021).

24 Wilko Graf von Hardenberg, "Knowing the Littoral: Perception and Representation of Terraqueous Spaces in a Global Perspective," *ISIS* 112, 1 (2021): 108–10; see also Philip E. Steinberg, "Of Other Seas: Metaphors and Materialities in Maritime Regions," *Atlantic Studies* 10, 2 (2013): 156–69. There has been an emergence of coastal history as distinct from maritime history, again with a strong cast from environmental history; I am indebted especially to John R. Gillis, *The Human Shore: Seacoasts in History* (University of Chicago Press, 2013); Christopher Pastore, *Between Land and Sea: The Atlantic Coast and the*

Transformation of New England (Harvard University Press, 2014); Isaac Land, "The Urban Amphibious," in *The New Coastal History: Cultural and Environmental Perspectives from Scotland and Beyond*, ed. David Worthington (Palgrave Macmillan, 2017); and a special Focus section on littorals in *ISIS* (112, 1 [2021]). In the (North) American context there has been a focus on southern shorelines, especially those in the paths of hurricanes, such as Paul Sutter and Paul M. Pressley, eds., *Coastal Nature, Coastal Culture: Environmental Histories of the Georgia Coast* (University of Georgia Press, 2018); and work by Ari Kelman and Craig Colten about Louisiana's Gulf coast.

25 Jennifer Bonnell and Sean Kheraj, "Urban Environmental History in Anglophone Canada: Omissions and Opportunities," *Urban History Review* 50, 1–2 (2022): 47.

26 Richard Judd, *Second Nature: An Environmental History of New England* (University of Massachusetts Press, 2014), 11–12. The evolution of the field in Atlantic Canada can be sketched in review essays such as Brian Payne, "The Environmental Historiography of the Maritime Peninsula," *Acadiensis: Journal of the History of the Atlantic Region* 45, 1 (2016): 163–77; and Blair Stein, "Environmental History at Work: New Environmental Histories of Canada and Atlantic Canada," *Acadiensis* 50, 1 (2022): 135–48.

27 Consider that the two most recent collections on the environmental history of the region contain just one essay on urban landscapes between them. Claire Campbell, Edward MacDonald, and Brian Payne, eds., *The Greater Gulf: Essays on the Environmental History of the Gulf of St. Lawrence* (McGill-Queen's University Press, 2019); and Claire Campbell and Robert Summerby-Murray, eds., *Land and Sea: Environmental History in Atlantic Canada* (Acadiensis Press, 2013).

28 From the older work by William Ganong (for New Brunswick) and Andrew Hill Clark (for Prince Edward Island) to more recent work by Stephen Hornsby, Joshua MacFadyen, and most prolifically, Graeme Wynn. See Wynn, "W.F. Ganong, A.H. Clark and the Historical Geography of Maritime Canada," *Acadiensis* 10, 2 (1981): 5–28; "Thinking about Mountains, Valleys and Solitudes: Historical Geography and the New Atlantic History," *Acadiensis* 31, 1 (2001): 129–45; "'Tracing One Warm Line Through a Land so Wide and Savage.' Fifty Years of Historical Geography in Canada," *Historical Geography* 40 (2012): 5–32.

29 See, for example, T.W. Acheson, "The National Policy and the Industrialization of the Maritimes, 1880–1910," *Acadiensis* 1, 2 (1972): 3–28; L.D. McCann, "Staples and the New Industrialism in the Growth of Post-Confederation Halifax," *Acadiensis* 8, 2 (1979); and Kris Inwood, ed., *Farm, Factory, and Fortune: New Studies in the Economic History of the Maritime Provinces* (Acadiensis Press, 1993).

30 Arthur Lower, *Unconventional Voyages* (Ryerson, 1953), viii–ix. For a fuller discussion of this point, see Claire Campbell, "'Rising with the Tide of History': The Age of Sail as Industrial Alibi," *Papers in Canadian History and Environment* 2, 1 (2019): 1–37.

31 John Reid, speaking at the Atlantic Canada Studies Conference, Orono, 2024, referring to Elizabeth Mancke's influential essay, "Spaces of Power in the Early Modern Northeast," in *New England and the Maritime Provinces: Connections and Comparisons*, ed. Stephen J. Hornsby and John G. Reid (McGill-Queen's University Press, 2005). I have been particularly influenced by the work of Stephen Hornsby, Jeffers Lennox, Christopher Pastore, Thomas Wickman, and Jack Bouchard.

32 Jacques-Nicholas Bellin, "Remarks by M. Bellin in Relation to His Maps Drawn for P. Charlevoix's History of New France," *The Gentleman's Magazine*, February 1746, 72; Jeffers Lennox, "An Empire on Paper: The Founding of Halifax and Conceptions of Imperial Space, 1744–55," *Canadian Historical Review* 88, 3 (2007): 400.

33 Ironically, the digital turn and the use of platforms like ArcGIS, which use Google maps as the default base map or end-state, have encouraged this assumption that digital maps are inherently correct and protected from human bias.

34 J.B. Harley, *The New Nature of Maps: Essays in the History of Cartography*, ed. Paul Laxton (Johns Hopkins University Press, 2001), 37.

35 Jill Desimini and Charles Waldheim, *Cartographic Grounds: Projecting the Landscape Imaginary* (Princeton Architectural Press, 2016), 11.

36 Matthew Edney, *Cartography: The Ideal and Its History* (University of Chicago Press, 2019), 77. There has been exciting scholarship on mapping and Indigenous, imperial, and early national identities and territories, but much of this sits in the dynamics of imperial or intellectual history, more than environmental history. Works that have informed the Canadian seaboard include The History of Cartography project from the University of Chicago Press; Stephen Hornsby, *Surveyors of Empire: Samuel Holland, J.F.W. Des Barres, and the Making of the Atlantic Neptune* (McGill-Queen's University Press, 2011); S. Max Edelson, *The New Map of Empire: How Britain Imagined America Before Independence* (Harvard University Press, 2017); Jeffers Lennox, *Homelands and Empires: Indigenous Spaces, Imperial Fictions, and Competition for Territory in Northeastern North America, 1690–1763* (University of Toronto Press, 2017); Chet Van Duzer and Lauren Beck, *Canada Before Confederation: Maps at the Exhibition* (Vernon Press, 2017); and Martin Brückner, ed., *Early American Cartographies* (Published for the Omohundro Institute of Early American History and Culture by the University of North Carolina Press, 2011) and *The Social Life of Maps in America, 1750–1860* (University of North Carolina Press, 2017).

37 Nancy Coulin and Carola Hein, "The North Sea: New Perspectives on the Sea-Land Continuum," in *The Urbanisation of the Sea: From Concepts and Analysis to Design* (nai010 Publishers, 2020), 12.

38 Adam Grydehøj, "Islands as Legible Geographies: Perceiving the Islandness of Kalaallit Nunaat (Greenland)," *Journal of Marine and Island Cultures* 7, 1 (2018): 1–11.

39 Edited historical atlases are particularly helpful here. Consider the work of Derek Hayes or the three-volume *Historical Atlas of Canada* series (1987–93).

40 Joana Gaspar de Freitas, Robert James, and Isaac Land, "Coastal Studies and Society: The Tipping Point," *Coastal Studies & Society* 1, 1 (2021): 3–9. I thank Jeremy Chow for our discussions on this.

41 Sven Beckert, Ulbe Bosma, Mindi Schneider, and Eric Vanhaute, "Commodity Frontiers and the Transformation of the Global Countryside: A Research Agenda," *Journal of Global History* 16, 3 (2021): 435–50.

42 *Great Ports of the Mercantile Era*, ed. Stephen J. Hornsby (University of Maine Press, 2023); also plates 36 and 45 in Stephen J. Hornsby and Richard W. Judd, eds., *Historical Atlas of Maine* (University of Maine Press, 2015).

43 Harvey, *The Limits to Capital*, 233; and "Globalization and the 'Spatial Fix,'" *Geographische revue* 3 (2001): 27.

44 Dunbar-Hester, *Oil Beach* (2023), 16. The Subcommission on Quaternary Stratigraphy, Anthropocene Working Group, has used the term "technofossil" to describe an Anthropogenic deposit. "What Is the Anthropocene? – Current Definition and Status" (2019).

45 Gillis suggests this in "Learning to Live with Coasts," *The Human Shore* (2013), 193. The idea of sunk capital – excess capital fixed in place – is part of Harvey's argument as to why capitalism has been able to entrench itself so successfully in what he called "the spatial fix."

It turns out there is much to be gained from focusing on coastlines in a critical analysis of modern capitalism.

46 The Charter of the City of Saint John (1785), in Thomas Wetmore files, S89-6, Saint John recorder's bound compilation, New Brunswick Museum; reprinted in *Canada's First City – Saint John, N.B.: The Charter of 1785 and Common Council Proceedings* (Lingely Printing, 1962); *The Royal Gazette and the New-Brunswick Advertiser,* 20 January 1795, 4. See chapter 1.

47 "Commander Robinson's Report," submitted with the *Journal of the House of Assembly of Newfoundland,* first session of the fourteenth assembly (1883), 124–207 of the Land Tenure report. See chapter 2.

48 Written at Observation Cove, Hillsborough Harbour, "Capt. Holland's account of the Island of St. John, in his letter of 6 Octr. 1765," to John Pownall, Secretary of the Board of Trade, Colonial Office records, National Archives; reproduced in Earle Lockerby and Douglas Sobey, *Samuel Holland: His Work and Legacy on Prince Edward Island* (Island Studies Press, 2015), 173–74. See chapter 3.

49 Hansard, 3 March 1905, House of Commons *Debates*, 10th Parliament, 1st Session, vol. 2, 2025.

50 Frederick William Cowie, *Report to the Honourable Frank Cochrane, Minister of Railways and Canals, on Halifax Harbour and the development of a project of modern ocean terminals* (Montreal, 1913), 49. See chapter 4.

51 Hugh MacLennan, *Barometer Rising,* New Canadian Library edition (McClelland & Stewart, 1989), 12.

52 Jacques Cartier, *Brief recit & fuccincte narration, de la nauigation faicte es yiles de Canada, Hochelage & Saguenay & autres* (A. Le Clerc/P. Roffet, 1545), available at Bibliothèque Nationale de France, https://gallica.bnf.fr/ark:/12148/bpt6k521342.

Chapter 1: Harbour Improved

1 Address from 9 August by Harvey at Saint John, enclosed to Lord Glenelg, 12 August 1838, Copies Or Extracts of Correspondence Relative to the Affairs of British North America, Colonial Office, 1839.

2 Maxwell Sutherland, "Bruce, Robert George," *Dictionary of Canadian Biography*, vol. 4 (University of Toronto/Université Laval, 2003–); and Stephen Hornsby, *Surveyors of Empire: Samuel Holland, J.F.W. Des Barres, and the Making of the Atlantic Neptune* (McGill-Queen's University Press, 2011), 33–34.

3 Bruce suggested that "the most advantageous situation for a Fort to Command this Harbour and River" was on the *other* side of the harbour from Fort Frederick, roughly where Queen's Square would later be laid out. Despite Bruce's recommendation, during the American Revolution Fort Howe would be built on higher ground and a bit deeper in the harbour. On the fall of Louisbourg, see A.J.B. Johnston, *Endgame, 1758: The Promise, the Glory, and the Despair of Louisbourg's Last Decade* (University of Nebraska Press, 2007).

4 *Census of the Canadas, 1851–2* (J. Lovell, 1853) shows Toronto at 30,775, Quebec at 61,526, and Montreal at 77,381. The *Population Return and Other Statistics of the Province of New Brunswick for 1851* lists Portland (8,429) and the city of Saint John (22,745) separately, plus the adjacent parishes of Lancaster, St. Martins, and Simonds in the County of Saint John. Appendix, *Journal of the House of Assembly of the Province of New Brunswick,* 7 January–7 April 1852.

5 T.W. Acheson, *Saint John: The Making of a Colonial Urban Community* (University of Toronto Press, 1985), 5.
6 Bailey Brogan, Jamylynn McDonald, Blaine Lyons, Peter Johnston, and Graeme Stewart-Robertson, "Appendix A: Coastal Hazard Flood Risk Mapping," in *Climate Change Adaptation Plan for Saint John* (Atlantic Coastal Action Program, 2020), 62–68.
7 Ged Martin, "Geography and Governance: The Problem of Saint John (New Brunswick) 1785–1927," https://www.gedmartin.net/martinalia-mainmenu-3/237-geography-and-governance-the-problem-of-saint-john-new-brunswick-1785-1927?.
8 Robert Cooney, *The Autobiography of a Wesleyan Methodist Missionary* (Montreal, 1856), 215.
9 David Bell, *Early Loyalist Saint John: The Origin of New Brunswick Politics, 1783–1786* (New Ireland Press, 1983), 47; part of a flurry of interest in Loyalist history approaching the bicentennial of the Loyalist arrival.
10 Any discussion of nineteenth-century New Brunswick must start with Graeme Wynn's foundational *Timber Colony: A Historical Geography of Early Nineteenth-Century New Brunswick* (University of Toronto Press, 1981).
11 A brief but useful introduction to the city's nineteenth-century history is C.M. Wallace, "Saint John, New Brunswick (1800–1900)," *Urban History Review* 4 (1975): 12–21. Elizabeth W. McGahan's exhaustive account of the port is indispensable but concerned primarily with the political dynamics *after* Confederation with Canada: *The Port of Saint John*, vol. 1, *From Confederation to Nationalization, 1867–1927: A Study in the Process of Integration* (National Harbours Board, 1982).
12 B.H. MacOwan, "The Evolution of a Regional Urban Network: New Brunswick and Nova Scotia, 1871–1971" (PhD thesis, University of Waterloo, 1986), 66.
13 George E. Fenety, *Political notes and observations, or, A glance at the leading measures that have been introduced and discussed in the House of Assembly of New Brunswick ...* (S.R. Miller, 1867), 72. There are innumerable versions of this story but most include a standard cast of characters: rocky shoreline, dense forest, cold winter.
14 Stephen Hornsby, *British Atlantic, American Frontier: Spaces of Power in Early Modern British America* (University Press of New England, 2005), 180–203; James D. Kornwolf and Georgiana Wallis Kornwolf, *Architecture and Town Planning in Colonial North America* (Johns Hopkins University Press, 2002), 3: 1330–74; and John Leroux et al., *Building New Brunswick: An Architectural History* (Goose Lane Editions, 2008), 44.
15 Jason Hall, "River of Three Peoples: An Environmental and Cultural History of the Wəlastəw/rivière St. Jean/St. John River, c. 1550–1850" (PhD diss., University of New Brunswick, 2015) and "Maliseet Cultivation and Climatic Resilience on the Wəlastəkw/St. John River During the Little Ice Age," *Acadiensis* 44, 2 (2015): 3–25; Michal A. Pawling, "Wəlastəkwey (Maliseet) Homeland: Waterscapes and Continuity Within the Lower St. John River Valley, 1784–1900," *Acadiensis* 46, 2 (2017): 5–34. Different Indigenous territories can be visualized at native-land.ca. The complicated seventeenth history of the site can be understood through two locations: Fort LaTour and Fort Charnisay. See Canada's Register of Historic Places (www.historicplaces.ca). The dykes are mentioned in *St. John and its business: A history of St. John and a statement in general terms of its various kinds of business successfully prosecuted* (1875), 13.
16 Béatrice Craig, *Backwoods Consumers and Homespun Capitalists: The Rise of a Market Culture in Eastern Canada* (University of Toronto Press, 2009), 27.
17 The Charter of the City of Saint John (1785), in Thomas Wetmore files, S89-6, Saint John recorder's bound compilation, New Brunswick Museum; reprinted in *Canada's First City* –

Saint John, N.B.: The Charter of 1785 and Common Council Proceedings (Lingely Printing, 1962). Also The charter of the city of Saint John in the province of New Brunswick 1811; and An Act for confirming unto the City of Saint John its Rights and Privileges [City Charter]) 1786. The charters generally defined means to maintain public order, regulate commerce, arrange government positions, etc.

18 See "Appendix I," *The historical charters and constitutional documents of the city of London* (Whiting, 1884), 132–38; *The charters of the city of London, which have been granted by the kings and queens of England since the conquest* (London: J. Hodges, 1745), 80–81, 112–13; Hall, "River of Three Peoples," 348. Writing in the early sixteenth century, Robert Fabyan recorded an episode from 1406 when the mayor and aldermen "as coseruatours of that ryuer" pulled up weirs on the Thames that had interfered with fish stocks. *The new chronicles of England and France, in two parts: by Robert Fabyan. Named by himself The concordance of histories*, ed. Henry Ellis (F.C. & J. Rivington, 1811) 572–73. The Crown had also granted standing of conservator to Newcastle-upon-Tyne by the mid-sixteenth century. As Leona Skelton observes, the responsibility meant primarily "preventing the choking up of what was effectively a great liquid highway that was crucial to trade," which – interestingly – already included large shipments of coal. *Tyne After Tyne: An Environmental History of a River's Battle for Protection 1529–2015* (White Horse Press, 2017), 35.

By the eighteenth century, "conservator" came to be understood as an arbiter, but also "A protector, preserver, or maintainer," as in Jacob Giles, *A new law-dictionary* (In the Savoy, 1732). See also John Noorthouck, *A New History of London Including Westminster and Southwark* (R. Baldwin, 1773); William Ferneley Allen, *The Corporation of London: Its Rights and Privileges* (1858); and George Norton, *Commentaries on the history, constitution, and chartered franchises of the city of London* (1869), 390–91. The conservator responsibility was formally reconfigured in 1857 as "the Conservators of the River Thames," subsequently the Thames Conservancy. See Jim Clifford, *West Ham and the River Lea: A Social and Environmental History of London's Industrialized Marshland, 1839–1914* (UBC Press, 2017). Acheson makes the distinction between conservator and developer (i.e., responsible for ferries, utilities), which may be a distinction without a difference (182).

In light of our questions of the Anthropocene, it is interesting that James I issued another charter shortly thereafter that dealt entirely with the city's historical right to measure and collect revenue from all coal – "call'd Stone-Coals, Pit-Coals, Earth-Coals, and all other Coals weighable" – brought by the Thames, at least in part to limit "the great Harm and Choaking up the Stream of the same River" by coal falling or dumped by smaller vessels. *Historical charters and constitutional documents*, 157; *Charters of the city of London* (1745), 100.

19 This was known as the Montgomerie Charter, after the incumbent governor of the colony. While Loyalists in New Brunswick may have been loathe to credit the new commercial capital of the "revolted states," London and New York were the most important estuarine metropolitan centres with which the Loyalist elite would have been familiar. *Laws and ordinances, ordained and established by the mayor, aldermen and commonalty of the city of New-York, in Common Council convened* (Samuel and John Loundon, 1786), 30; Hendrik Hartog, *Public Property and Private Power: The Corporation of the City of New York in American Law, 1730–1870* (University of North Carolina Press, 1983), 28, 34, and 53–54. My thanks to Kara Schlichting for her suggestions on this.

20 Contrary to his earlier claim that nature had been "unkind," Bell notes that "through their control of the harbour and the corporation lands, the councillors possessed the richest natural resources in the province, including not only the harbour, its wharves, water lots,

tidal power, flats, and water front, but also the most valuable salmon fishery on the Atlantic coast" (*Early Loyalist Saint John*, 42).

21 Peter Coates and James Williams offer useful reflections on the concept of improvement in *The Illusory Boundary: Environment and Technology in History*, ed. Martin Reuss and Stephen H. Cutcliffe (University of Virginia Press, 2010).

22 In June 1792, for example, the council passed a statute "to prevent the taking gravel, sand, stones, and shingles from certain parts of the Beaches of the Harbour." See *Canada's First City – Saint John, N.B.* (1962), 186.

23 "Miscellaneous remarks and observations on Nova Scotia, New Brunswick and Cape Breton, Supposed to be written by the Surveyor General of Nova Scotia," *Collections of the Massachusetts Historical Society*, March 1794, 100. My thanks to Arn Keeling for his thoughts on this point.

24 C.M. Wallace, "Saint John Boosters and the Railroads in Mid-nineteenth Century," *Acadiensis* 6, 1 (1976): 73. On the geography of hinterlands, see Wynn, *Timber Colony;* MacOwan, "Evolution of a Regional Urban Network," 70–72; this is Martin's main concern in "Geography and Governance."

25 *The Royal Gazette and the New-Brunswick Advertiser,* 20 January 1795, 4. The poem, "On Opening a Little Theatre in This City," was performed at a local tavern. I am grateful to Gwendolyn Davies for discussing this with me. See her article "Consolation to Distress: Loyalist Literary Activity in the Maritimes," *Acadiensis* 16, 2 (1987): 58; and Foreword, *The Creative City of Saint John,* ed. Gwendolyn Davies, Peter Larocque, and Christl Verduyn (Formac Publishing, 2018), 9.

26 See Martyn J. Bowden, *Great Ports of the Mercantile Era*, ed. Stephen J. Hornsby (University of Maine Press, 2023), and plate 36 in Stephen J. Hornsby and Richard W. Judd, eds., *Historical Atlas of Maine* (University of Maine Press, 2015).

27 Moses H. Perley, "On the Early History of New Brunswick: A portion of a lecture delivered before the Mechanics' Institute, St. John, in 1841 ... with notes by W.F. Ganong," 1891, 20–21.

28 An Act to provide for the Support of a Light House, to be built upon Partridge Island (1788); An Act to extend the powers of the Mayor, Aldermen, and Commonalty, of the City of Saint John, for preventing the incumbering and filling up of the Harbour of the said City, to the Waters and Places of Anchorage in the Vicinity thereof (1824). Charter status notwithstanding, the city would never be entirely financially solvent; the colonial assembly regularly considered legislation related to and funding for activities on the river and harbour. Laws to prevent or remove "encumbrances" for the movement of fish were repeatedly (and evidently ineffectually) passed in post-Revolutionary British North America. I am indebted to the British North American Legislative Database (https://bnald.lib.unb.ca) for digitizing and transcribing these pieces of legislation.

29 The council sought to ensure stairs at the end of any street leading to the water but discovered March was an inopportune time with "such a body of ice & snow on the ground that they cannot at present with any degree of certainty ascertain the expense nor the most convenient place to affix them," *Canada's First City,* 25 March 1790, 148.

30 Acheson, *Saint John*, 31.

31 Roger Sarty and Doug Knight, *Saint John Fortifications, 1630–1956* (Goose Lane Editions, 2003). In his 1848 *Plan of Military or Government Land, Lower Cove, Saint John, N.B.* (New Brunswick Museum), Thomas O'Keleher argued for the need to have this border clearly

and correctly marked so that "no Private Individual" should build to within a foot of the government land.

32 Abraham Gesner, *New Brunswick; with notes for emigrants* (London: Simmonds & Ward, 1847), 121; and J.F.W. Johnston, *Notes on North America* (London: Blackwood and Sons, 1851), vol. i, 39.

33 Anthony Lockwood, *Report on the projected canal across the istmus [sic] that divides Nova-Scotia and New-Brunswick: explored and levelled in the autumn of 1819* (n.p.). The boosterish *St. John and Its Business* (1875) says the first proposal was mooted about 1800 (28).

34 An Act for the Incorporation of the Saint John Mills and Canal Company (1834); Wynn, *Timber Colony*, 103–4. The canal is now filled in and buried under an Irving pulp plant.

35 *Report of the commissioners appointed to examine into the present condition of the harbour of Saint John, and to report the same to His Excellency the Lieutenant-Governor of the province: with various documents relating thereto* (1839), 1–2.

36 Submission by Lauchlan Donaldson, *Report of the commissioners*, 9–10.

37 An Act to limit the extent and regulate the building of Wharves on the eastern side of the Harbour of Saint John (1840).

38 D.H. Waterbury, "Retrospective Ramble over Historic St. John," Collections of the New Brunswick Historical Society #10 (1919), 87; and John Willet, "Epitomised History of Saint John," Collections of the New Brunswick Historical Society #11 (1927), 177.

39 An Act to make more effectual provision for repairing the Aboideau [sic] or Bridge across the Marsh Creek in the City and County of Saint John (1808); An Act to provide for the drainage and protection of the Great Marsh in the City of Saint John, and the Parishes of Portland and Simonds in the City and County of Saint John, and the maintenance of an Aboideau across the Marsh Creek near the City of Sai (1862). A parallel act was directed at the parish of Lancaster southwest of the city in 1864. The phrasing "aboideau or bridge" suggests the bridge was built atop an older Acadian earthwork. For a wonderful overview of the Acadian landscape, see the first chapter of Ronald Rudin, *Against the Tides: Reshaping Landscape and Community in Canada's Maritime Marshlands* (UBC Press, 2021).

Filling wetland was partly a property investment and partly a standard practice of public health. The best illustration of how the two conveniently aligned was at the Mill Pond between Portland and Saint John, which – as a valley – suited the prospect of a railway line into the city but then raised concern during a cholera epidemic in 1854.

40 Peter Fisher, *Sketches of New-Brunswick: containing an account of the first settlement of the province: with a brief description of the country, climate, productions, inhabitants, government, rivers, towns, settlements, public institutions, trade, revenue, population &c.* (1825), 33.

41 Peter Fisher, *Notitia of New-Brunswick, for 1836, and Extending Into 1837* (London: H. Chubb, 1838), 66. "Emporium of New Brunswick" is repeated in other contemporary accounts, and the epithet "Liverpool of America" appears in promotional material for decades. On Liverpool's own port history, see the chapter in Peter Coates, *A Story of Six Rivers: History, Culture and Ecology* (Reaktion Books, 2013).

42 This map of Saint John was part of a survey of the Bay of Fundy for the British Admiralty organized by Captain (soon to be rear-Admiral) William F. Owen, which included a series of tide-gauges to measure tides as accurately as possible. Owen had taken up an inherited title to nearby Campobello Island before moving to Saint John and was an active participant in local affairs and governance, advocating for industrial development. Paul G. Cornell, "Owen, William Fitz William," *Dictionary of Canadian Biography*, vol. 8 (University of Toronto/Université Laval, 2003–).

43 An Act to encourage the erection of a Passage Boat to be worked by Steam for facilitating the communication between the City of Saint John and Fredericton (1812). Subsequent legislation required the contractors "to provide for the greater safety of passengers" with adequate life boats, fire buckets, and navigational lights, to mitigate accidents and injury "by the exploding, collapsing, blowing up, or giving way of any part of the machinery."

44 An Act to provide for the erecting and making certain Wharves and improvements in the Harbour of Saint John (1852).

45 Montreal *Gazette*, 13 June 1833, 2. The New Brunswick legislature was subsidizing steam traffic between Saint John and Halifax by the 1830s.

46 Gesner, *First Report on the Geological Survey of New Brunswick* (H. Chubb, 1839), 78; Fisher, *Notitia of New-Brunswick*, 13. After a survey of mineral resources in Nova Scotia, Gesner would be contracted to undertake a survey of New Brunswick and subsequently Prince Edward Island. He later became famous for inventing "coal-oil," or kerosene.

In fact, capital followed resource, and more momentum went toward the larger seams in Nova Scotia. Scholarship likewise follows historical activity, so there is a substantial body of scholarship on the labour and management of coal mining in Nova Scotia but very little on coal mining in New Brunswick, apart from a brief discussion of the Grand Lake district in Allen Seager, "Minto, New Brunswick: A Study in Canadian Class Relations Between the Wars," *Labour/Le Travail* 5 (1980): 81–132. Andrew Watson offers an overview of "Coal in Canada," in *Powering Up Canada: The History of Power, Fuel, and Energy from 1600* (McGill-Queen's University Press, 2016), but compared to larger fields elsewhere in Canada those in New Brunswick are too small to register.

47 Wynn, *Timber Colony*, 109.

48 Thomas Leavitt and N.S. Demill, "Report on the effect of sawdust on the harbour of Saint John," *Journal of the House of Assembly*, 1849, Appendix.

49 An Act for more effectually securing the Navigation of the River and Harbour of Saint John, in the City and County of Saint John (1844); An Act in addition to and in amendment of the Act relating to the Navigation of the River and Harbour of Saint John (1849); An Act relating to the Navigation of the River and Harbour of Saint John (1851), which exempted the Carleton Tide Mills entirely because the act would be "ruinous."

50 Thomas H. Keefer, *Report upon the harbor of Saint John, New Brunswick* (1861), 4.

51 Keefer, *Report upon the harbor*, 3, 5, 13, 15.

52 McGahan, *Port of Saint John*, 79; Robert Babcock, "Re-examining the Economic Underdevelopment of the Maritime Provinces: A Case Study of Portland, Maine, and Saint John, New Brunswick," in *New England and the Maritime Provinces: Connections and Comparisons*, ed. Stephen Hornsby and John Reid (McGill-Queen's University Press, 2005).

53 Eric Sager and Gerald E. Panting, *Maritime Capital: The Shipping Industry in Atlantic Canada, 1820–1914* (McGill-Queen's University Press, 1990), 160; Charles Forward, "The Development of Canada's Five Leading National Ports," *Urban History Review/Revue d'histoire Urbaine* 10, 3 (1982): 25–45.

54 *Saint John as a Canadian winter port and terminus of the Canadian Pacific and Intercolonial Railways* (City Corporation/Board of Trade, 1898).

55 Bowden, "Industrial Portland," Plate 45, in *Historical Atlas of Maine* (2015). Carola Hein and Dick Schubert make this observation about the effect of the turn to containerization in "Introduction: Resilience, Disaster, and Rebuilding in Modern Port Cities," in "Resilience in Seaports," special issue, *Journal of Urban History* 47, 2 (2021): 235–49.

56 Ian Sclanders, "Saint John," in Frederick William Wallace, *The Romance of a Great Port* (Barnes, 1935), 8. The history professor would add that the use of the passive voice is part of the problem.

Chapter 2: Limits Defied

1 Louis Legrand Noble, *After icebergs with a painter: A summer voyage to Labrador and around Newfoundland* (New York: D. Appleton, 1861) 31, 37.

2 William Fraser Rae, *Newfoundland to Manitoba, through Canada's maritime, mining, and prairie provinces* (G.P. Putnam's Sons, 1881), 50.

3 Norman Catto, Olav Slaymaker, and Dori J. Kovanen, "Human-Modified Landscapes," in *Landscapes and Landforms of Eastern Canada,* ed. Olav Slaymaker and Norman Catto (Springer, 2020), 501. See also the various research reports by archaeologist Peter Pope, such as *St. John's Harbour Area Archaeological Potential* (Past Present, Historic Sites and Material Culture Consulting, 1991); *Under Water Street: The Early Waterfront of St. John's, Newfoundland* (ICHAM Publications, 1996); *Summer 2000 Excavations at 327 Water St., St. John's (CjAe-08)* (St. John's Archaeology Project, unpublished report, 2003).

4 Peter Pope, *Fish into Wine: The Newfoundland Plantation in the Seventeenth Century* (Omohundro Institute of Early American History and Culture/University of North Carolina Press, 2004). Newfoundland, as Ktaqmkuk, remains the unceded territory of the Mi'kmaq, with the earlier Beothuk people wiped out by the early nineteenth century.

5 Olaf Janzen, "The Royal Navy and the Defence of Newfoundland During the American Revolution," *Acadiensis* 14, 1 (1984): 28–48, reprinted in *War and Trade in Eighteenth-Century Newfoundland* (Liverpool University Press, 2013), 193–214; and Hannah M. Kaemmer, "Expertise and Empire: Fortification Building and the English Ordnance Office, 1660–1714" (PhD diss., Harvard University, 2024).

6 Jean M. Murray, ed., *The Newfoundland Journal of Aaron Thomas, Able Seaman in H.M.S. Boston* (Longman's Green, 1968), 51–52.

7 The mercantile shore is described in John J. Mannion, "St. John's," Plate 27 in R. Cole Harris, ed., *Historical Atlas of Canada,* vol. 1, *From the Beginning to 1800* (University of Toronto Press, 1987); Catto et al., "Human-Modified Landscapes," 495; Norman R. Catto and L. St. Croix, "Urban Geology of St. John's, Newfoundland," in *Urban Geology of Canadian Cities,* ed. Paul F. Karrow and Owen L. White (Geological Association of Canada, 1998), 449.

8 George A. Nader, *Cities of Canada,* vol. 2, *Profiles of Fifteen Metropolitan Centres* (Macmillan of Canada, 1975–76), 14.

9 Jerry Bannister, *The Rule of the Admirals: Law, Custom, and Naval Government in Newfoundland, 1699–1832* (Osgoode Society for Canadian Legal History/University of Toronto Press, 2003).

10 Governor Gower to Earl Camden, 19 November 1804, *Despatches from the Governor of Newfoundland to the Secretary of State [for the Colonies], vol. 1, 1800–1813,* Memorial University Archives.

11 For a detailed discussion of the economic growth of this period, see Shannon Ryan, "Fishery to Colony: A Newfoundland Watershed, 1793–1815," *Acadiensis* 12, 2 (1983): 34–52.

12 "An act for taking away the public use of certain ships rooms in the town of Saint John['s], in the island of Newfoundland, and for instituting Surrogate Courts on the coast of Labrador, and in certain islands adjacent thereto," 31 May 1811, Georgii III. Regis. Cap.

XLV. On the longer political history, see, for the earlier part of the nineteenth century, Bannister, *Rule of the Admirals* and Paul O'Neill, *The Oldest City: The Story of St. John's, Newfoundland* (Press Porcepic, 1975); and for the later nineteenth and twentieth century, Raymond Blake and Melvin Baker, *Where Once They Stood: Newfoundland's Rocky Road Towards Confederation* (University of Regina Press, 2019).

13 Governor R.G. Keats to the Earl of Bathurst, 13 December 1813, *Despatches from the Governor of Newfoundland to the Secretary of State [for the Colonies], vol. 1, 1800–1813* and Colonial Office papers CO 194, vol. 54 (1813), no. 159 MUN/DAI.

14 Lewis Anspach, *A history of the island of Newfoundland, containing a description of the island, the banks, the fisheries and trade of Newfoundland, and the coast of Labrador* (London, 1819), 254–55.

15 Jenkin Jones, Secretary, the Phoenix Fire [Insurance Co.] Office, London, to Thomas Lack, Office of the Committee of the Privy Council for Trade, 15 December 1817, in *Despatches received by the Governor of Newfoundland from the Secretary of State for the Colonies, vol. 2, 1814–1819*, 120; Charles Pedley, *The history of Newfoundland from the earliest times to the year 1860* (Longman, Roberts & Green, 1863), 417. Fire was a familiar and defining feature of cities in both the New World and Old, so the language here is interesting.

16 John Oldfield, "Memoranda on Newfoundland during a residence from October 25, 1830 to October 12, 1835," Centre for Newfoundland Studies, Memorial University.

17 Benjamin L. Carp, "Fire of Liberty: Firefighters, Urban Voluntary Culture, and the Revolutionary Movement," *The William and Mary Quarterly* 58, 4 (2001): 781–818; and Cindy Ermus, *Urban Disasters* (Cambridge University Press, 2023). Fires loom large in the history of American cities, especially those in Chicago (1871) and San Francisco (1906), but the St. John's fire story is notable for its repetitive genealogy with fires of record spanning almost the entire century.

18 Some of the more notable fires took place in 1816, 1817, 1819, 1833, 1839, 1846, 1879, and 1892. See Paul Butler, *St. John's, City of Fire* (Flanker Press, 2007); Philip Tocque, "The great fires at St. John's: dates back to 1816," *Evening Telegram*, 3 September 1892, 2.

19 Jones, Phoenix Fire [Insurance Co.] Office, London, 15 December 1817, 120.

20 An Act to provide for the Management of a certain Public Wharf and Landing in the Town of Saint John's (12 June 1834). See also "An Act to regulate the Building of Houses in Water-Street in the Town of Saint John's" (1 August 1833) and "An Act to regulate the Re-building of the Town of St. John's, and the Drainage and Sewerage of the same, and to Repeal certain Acts therein mentioned" (4 August 1846). The expansion of water supply after the 1846 fire unfortunately led to ludicrous claims that the city was now inured to future disaster. See *History of Newfoundland from the earliest times*, 420; Joseph Hatton and M. Harvey, *Newfoundland: the oldest British colony, its history, its present condition, and its prospects in the future* (1883), 110; and D.W. Prowse, *A history of Newfoundland from the English, colonial, and foreign records* (1895), 493.

21 Governor LeMarchant to Earl Grey, 24 June 1847, in *Copies or extracts of the correspondence between the Governor of Newfoundland and the Secretary of State for the Colonies, in reference to the appropriation of the subscriptions raised for the relief of the sufferers at St. John's by the fire in 1846*, 79; *Journal of the House of Assembly of Newfoundland 1850 (2nd Session)*, 140–43. The promenade was a contentious project among several where relief funds were suspected of being used for projects decidedly peripheral to fire relief or otherwise usurping city priorities; see, for example, "Marine Promenade," *Patriot*, 16 September 1847, 1; "The Useful and the Ornamental," *Morning Courier*, 13 October 1847, 2.

22 *A Sketch of the state of affairs in Newfoundland By A Late Resident in that Colony* (London: Sauders & Otley, 1841), 21.

23 Moses Harvey, *Text-book of Newfoundland history, for the use of schools and academies* (Boston, Doyle and Whittle, 1885).

24 "An Act to provide for the Management of the Municipal Affairs of the Town of St. John's" [Municipal Act], 1888; Melvin Baker, "The Influence of Absentee Landlordism on the Development of Municipal Government in 19th Century St. John's," *Newfoundland Quarterly* 81, 2 (1985): 19–25 and "The Incorporation of St. John's in 1888," in *Aspects of Nineteenth Century St. John's Municipal History* (H. Cuff Publications, 1982). By way of comparison, the other dates of incorporation include 1785 (Saint John), 1841 (Halifax), and 1855 (Charlottetown).

25 James R. Osgood & Co., *The maritime provinces: a handbook for travellers* (Boston: J.R. Osgood, 1875), 190; and *Colton's Illustrated Cabinet Atlas and Descriptive Geography* (New York, 1859), 43. Or as Eliot Warburton put it in *Hochelaga: Or, England In the New World* (1846), 19: "We find other cities coupled with words which at once give their predominant characteristic: London the richest, Paris the gayest, St. Petersburg the coldest. In one respect the chief town of Newfoundland has, I believe, no rival; we may, therefore, call it the fishiest of modern capitals."

26 See Šimková, *Urban Archipelago: An Environmental History of the Boston Harbor Islands* (University of Massachusetts Press, 2021).

27 Gerald Penney Associates, and Newfoundland and Labrador Provincial Archaeology Office, *St. John's Harbour Interceptor Sewer Archaeological Investigations 2004–2010: Technical Report* (Gerald Penney Associates, 2011); Peter Pope, *St. John's Harbour Area Archaeological Potential, St. John's Archaeological Inventory and Analysis, Phase I, October 1991* (Past Present, Historic Sites and Material Culture Consulting, 1991), 77.

28 "Commander Robinson's Report," submitted with the *Journal of the House of Assembly of Newfoundland,* first session of the fourteenth assembly (1883), 124–207 of the Land Tenure report. Unfortunately, the biographical information on Robinson is brief; see the *Encyclopedia of Newfoundland and Labrador*, vol. 4, 612.

29 "Commander Robinson's Report," 138–39, 133–34. On "wharving in" and other technologies and practices of making waterfront, see Nancy Seasholes's detailed and compelling *Gaining Ground: A History of Landmaking in Boston* (MIT Press, 2003).

30 "A table shewing the estimated areas that have been built out into the Harbour" and "Total area of encroachments on Harbour," "Commander Robinson's Report," 198–207. The harbourmaster consulted a number of maps, including an "Admiralty plan from 1813" (likely by Thomas Eaststaff) as well as those by Orlebar, Noad, and the Roe Brothers company.

31 An observation made by Bowden about the "mercantile triangle" where the base of the triangle stretches further and further along.

32 George Robinson, "Report of the Harbour Master, 1878," Appendix, *Journal of the House of Assembly of Newfoundland* (1879), 105, 389–90; "Report of the Harbour Master for 1879," in *Journal of the House of Assembly of Newfoundland* (1880), 671–76. Similar concerns had been expressed in Saint John, N.B.

33 "Report on the Anchorage in St. John's Harbour," in "Commander Robinson's Report," 151; Robinson, "Report of Harbor Master, 1879," 675.

34 "Land Tenure: The Condition of the Harbor," *Evening Mercury*, 25 June 1883, 4.

35 Baker, "Influence of Absentee Landlordism," 23.

36 "Report of Committee on Land Tenure," 13 May 1882, submitted with the *Journal of the House of Assembly of Newfoundland*, first session of the fourteenth assembly (1883), 6. The land tenure committee was made up of several of the most prominent actors in either the Legislative Council or House of Assembly and, more to the point, the commercial activity of the city waterfront, including the chair, Pierre Tessier, the premier, William Whiteway, and James Fox, Alexander Mackay, Ambrose Shea, Thomas Talbot, Joseph Little, and Robert Kent.

Tessier, Whiteway, Fox, Mackay, and Kent are listed as lessees on the south side of Water Street; Shea a freeholder and lessee. Tessier also testified.

37 Laurence F. Geran, 49; Robert H. Prowse, 73; Robert Hunt, 113, in "Report of Committee on Land Tenure," 6.

38 "Commander Robinson's Report," 141, 145–46. Michael Rawson discusses the theory of "tidal scour," engineering, and Boston's efforts to restore and enhance water flow in *Eden on the Charles: The Making of Boston* (Harvard University Press, 2010), 179–232. Tessier and others had already petitioned the assembly for dredging of the upper harbour, *Journal of the House of Assembly*, fourth session of thirteenth general assembly (1882), 153.

39 "Commander Robinson's Report," 147.

40 Cyrus Field, "Statement of some of the advantages attendant upon making St. John's, Newfoundland a port of call for transatlantic steamers London" (n.p., 1856). The colonial government had attempted to subsidize steamer service from the island, since the 1830s, to the mainland (Halifax), then American and British ports like Boston and Liverpool. See "An Act to ... re-appropriate the same for the purpose of facilitating a communication by Steam between the Port of St. John's and the Port of Halifax in the Province of Nova Scotia" (12 October 1839); "An Act for the Encouragement of Communication by Steam between certain Ports and St. John's" (14 June 1852).

41 Joseph Hatton and Moses Harvey, *Newfoundland: The oldest British colony, its history, its present condition, and its prospects in the future* (London: Chapman and Hall, 1883), 149–50. Hatton reported that the dock was "to be excavated out of the rocky boundary of the south side of the harbour, so that it will be literally hewn out of the solid rock, and, when once constructed, will last for all time."

42 There had been a small milling complex at this location since 1827. The Goad insurance map of 1880, or the time of Robinson's investigation, shows a bakery and foundry in the area, as well as the gasworks and a hospital.

43 "Report of Committee on Land Tenure," 34.

44 [Walter] Kinipple and [William] Morris, "Report on Proposed Graving Dock and Harbour Improvements, at St. John's, Newfoundland," 13 September 1878, Appendix 27, *Journal of the Legislative Council*, first session of the thirteenth assembly (1879), 405, 408. Kinipple had been involved with construction on docks at London, Greenock [Scotland], Quebec, and Esquimalt; Kinipple and Morris had been commissioned in 1874 for a study of the St. John's waterworks, and Morris directed the survey for a line of railway from St. John to Harbour Grace in 1880. Obituaries for Morris, *Minutes of the Proceedings of the Institution of Civil Engineers* 87 (1887): 422–23; and Kinipple, *Minutes of the Proceedings of the Institution of Civil Engineers* 150 (1902): 436–41.

45 Commander George Robinson, 20 December 1882, "A Return on the Docks in Boston and New York, Philadelphia and Baltimore," Appendix, *Journal of the House of Assembly* (1883). He also issued a separate report on borings at the head of the harbour in anticipation of

drilling footings for a dock. Robinson was critical of the condition of the dock at Boston, now thirty years old, the oldest such dock built and operated by Simpson & Co. (1854).

Sealing the dock would be of concern; engineer W.J. Croasdale had observed that "the rock on this Island is so literally bestrewen [sic] with crevises [sic] that it is utterly impossible to tell at what moment the works will be flooded with water." Report of W.J. Croasdale, Government Engineer, on the proposed dry dock, 25 April 1882, *Journal of the House of Assembly* (1882), 649.

46 "Report of the Harbour Master, 1879," Appendix, *Journal of the Legislative Council* (1880): 105; "Dry Dock Evidence before Select Committee of the Council," *Evening Telegram*, 25 May 1882, 2; and "Commander Robinson's Report," 140, 145. The Hansard only begins printing in 1909, so assembly and council debates are found in the St. John's newspapers. There had been a floating dry dock anchored off the south side, but this was deemed insufficient.

47 Earlier in 1882, the House of Assembly had granted a charter for a dry dock as part of a short-lived railway project, but the contract was transferred to the Simpson & Co. bid. The contract specified the dock would include "Pumping Machinery to be of sufficient capacity to remove water from the Dock, when occupied by a Steamship of Two Thousand Tons, in two and a half hours to dredge a Channel, four hundred feet in width, from end of Piers" – measured by charts from Harbour Master Robinson. "An Act to provide for the construction of a dry, or graving, dock" (21 April 1883), Statutes of Newfoundland. The dock was rebuilt in concrete in 1926.

48 And "supposing 'the Dock' turns out to be 'all that our fancy painted it,'" the *Telegram* asked, "the next question is, Will it pay? Will this other 'great civilising enterprise' prove to be of any real utility to us, and to what extent will it be useful?" "A Few More Observations about the dock. Not a Very Good 'Bargain,' After All," *Evening Telegram*, 3 April 1883, 1; "Mr. Simpson's Advent Amongst Us," *Evening Telegram*, 10 May 1883; "Completion of the Dry Dock," *Evening Telegram*, 5 December 1884, 4.

49 Herbert C. Burchell "The colonial government dry dock, St. John's Newfoundland," *Transactions of the Canadian Society of Civil Engineers* (1887). Another description of the construction of the St. John's yard can be found in C.H. Stockton's report, "Simpson's Timber Dry-Docks," *Proceedings of the United States Naval Institute* 13, 2 (1887). George Robinson had also shared a description of the dock with the Institution of Civil Engineers in London as he thought his colleagues would be interested in the design and execution of a "wood dock on a solid foundation." "Nov/1884 Graving Dock at St. John's Newfoundland, Plans of the Harbour by George Robinson," 14 November 1884, Institution of Civil Engineers library.

50 "Premier speech on dock Bill," 29 March 1883, *Newfoundland Patriot*, 25 July 1883, 1. The moniker of "Apostle of Progress" came from the *Evening Telegram*. J.K. Hiller, "Sir William Vallance Whiteway," *Dictionary of Canadian Biography*, vol. 8 (University of Toronto/Université Laval, 2003–).

51 David Alexander, "Newfoundland's Traditional Economy and Development to 1934," in *Newfoundland in the Nineteenth and Twentieth Centuries: Essays in Interpretation*, ed. James Hiller and Peter Neary (University of Toronto Press, 1980). Attempts at agriculture in the interior turned quickly to large-scale resource extraction and processing (coal, pulp and paper).

52 James Hiller, "The Railway and Local Politics in Newfoundland, 1870–1901," in Hiller and Neary, *Newfoundland in the Nineteenth and Twentieth Centuries*. Construction had begun in 1881, but the railway was only completed across the island in 1898. The operation of the dry dock was folded into the Reid conglomerate's operation of the railway with the infamous "Reid Contract" of 1898.

53 Melvin Baker, "Moses Monroe," *Dictionary of Canadian Biography*.

54 House of Assembly debates, 13 March 1893, *Daily Tribune*, 1 April 1893, 3.

55 "An act respecting the lands in the limits of the town of St. John's" of 1892 (The Land Tenure Act) deals more with the provision for longer leases and the provision for compensating tenants for "improvements" made for any lease shorter than ninety-nine years. The Tenants' League was formed two weeks after the Great Fire. See *Journal of the House of Assembly of Newfoundland*, fourth session of the sixteenth assembly, 15 August and 19 August 1892; *Journal of the House of Assembly of Newfoundland*, fifth session of the sixteenth assembly, 26 April 1893.

56 House of Assembly from 23–24 August 1892, reprinted in *Evening Telegram*, 16–22 September 1892. Morris (who considered Robinson's "the most complete report that has ever been laid before this House," 19 September) argued for the entire waterfront to be administered by a harbour trust, akin to today's Port Authority. St. John's came under the authority of the National Harbour Board in 1964 (following Confederation with Canada).

57 The 1874 census showed 30,574; the 1901 census, 29,594, counting the south side.

58 Isaac C. Morris, *The dawn of the twentieth century, as pertaining to Newfoundland* (St. John's: G.S. Milligan, 1899), 24.

59 Consider the poignant reflection on this geography by Yangdong Li, "Lighting and Temperature in St. John's, Newfoundland," 2002, https://niche-canada.org/2022/09/08/lighting-and-temperature-in-the-st-johns/. For some of the physical hazards left by this era of road construction in particular, see David Liverman, Norm Catto, and Martin Batterson, "Geological Hazards and Disasters in St. John's," *Newfoundland & Labrador Studies* 21, 1 (2006): 71–96.

60 Fiona Black, "The Politics of Energy Sources," in *The Democracy Cookbook: Recipes to Renew Governance in Newfoundland and Labrador*, ed. Alex Marland and Lisa Moore (Memorial University Press, 2017), 313. I make a similar observation of Nova Scotia and references to the age of sail in "'Rising with the Tide of History': The Age of Sail as Industrial Alibi." *Papers in Canadian History and Environment* 2, 1 (2019): 1–37.

See also Fiona Polack and Danine Farquharson, eds., *Cold Water Oil: Offshore Petroleum Cultures* (Routledge, 2022); and Sheena Wilson, Adam Carlson, and Imre Szeman, eds., *Petrocultures: Oil, Politics, Culture* (McGill-Queen's University Press, 2017), especially Tim Kaposy, "Petroleum's Longue Durée: Writing Oil's Temporalities into History," 389–406. Today, about a third of the province's revenues come from mining and fuel extraction.

61 Jerry Bannister, "A River Runs Through It: Churchill Falls and the End of Newfoundland History," *Acadiensis* 41, 1 (2012): 212. We should, Bannister warns, "be extremely wary of megaprojects in general and predictions of future power costs in particular. It should also have taught us to be wary of those who advocate a future unconstrained by the past" (224).

62 Tom Kierans files, Memorial University, Archives #14-048 and QEII Library Map Collection.

63 Edward Morris speaking at the House of Assembly, 23 August 1892, *Evening Telegram*, 17 September 1892, 2. My thanks to Mark McLaughlin for the observation about Signal Hill.

Chapter 3: Nature Opposed

1 Though there is little in the way of environmental history about Atlantic Canadian cities, water *does* figure prominently in essays about health and sanitation in Douglas Baldwin and Thomas Spira's edited collection, *Gaslights, Epidemics, and Vagabond Cows: Charlottetown in the Victorian Era* (Ragweed Press, 1988). Kate Scarth considers the cities that author L.M. Montgomery knew (Charlottetown and Summerside), focusing on parks and urban beautification, in "Born for City Life? L.M. Montgomery's Urban Prince Edward Island," *Island Magazine* (2020): 50–60. More generally, Peter Rider's *Charlottetown: A History* (Canadian Museum of Civilization/PEI Museum and Heritage Foundation, 2009) is a good starting point. Sailstrait, a blog by Harry Holman, is a rich archive of the history of the city and environs, especially its marine and maritime heritage: https://sailstrait.wordpress.com.

2 It is the only provincial capital to not rank in the top hundred largest cities in Canada (Census of Canada, 2021). By way of census bookends for the period covered in this chapter, we can chart a growth in population in Charlottetown from 407 (1798) to 11,203 (1911), and a slight increase from 9.3 percent to 12 percent of the Island's population. Data taken from *A return of the inhabitants ... 1798,* in Duncan Campbell, *History of Prince Edward Island* (Charlottetown: Bremner Bros., 1875) and the decennial census, Government of Canada.

3 Edward MacDonald, *If You're Stronghearted: Prince Edward Island in the Twentieth Century* (PEI Museum and Heritage Foundation, 2000), 226.

4 Other small- and medium-sized cities were making similar choices in the railway era, though the examples in Canada that have been documented by historians – Hamilton and Toronto, Ontario – are located on the Great Lakes; significant water bodies to be sure, but not oceanic. See Thomas McIlwraith, "Digging Out and Filling In: Making Land on the Toronto Waterfront in the 1850s," *Urban History Review/Revue D'histoire Urbaine* 20, 1 (1991): 15–33; and Nancy Bouchier and Ken Cruikshank, *The People and the Bay: A Social and Environmental History of Hamilton Harbour* (UBC Press, 2016).

5 "Written at Observation Cove, Hillsborough Harbour, 'Capt. Holland's account of the Island of St. John, in his letter of 6 Octr. 1765,'" to John Pownall, Secretary of the Board of Trade, Colonial Office records, National Archives; reproduced in Earle Lockerby and Douglas Sobey, *Samuel Holland: His Work and Legacy on Prince Edward Island* (Island Studies Press, 2015), 173–74. On the layered history of the site in public interpretation, see https://www.pc.gc.ca/en/lhn-nhs/pe/skmaqn.

Holland was commissioned in 1762 to undertake a general survey of the "Northern District" (north of Virginia) of the newly British North America. The three "County Towns" were Princetown on Malpeque Bay, Charlottetown on Hillsborough Bay, and Georgetown on Cardigan Bay. See Stephen Hornsby, *Surveyors of Empire: Samuel Holland, J.F.W. Des Barres, and the Making of the Atlantic Neptune* (McGill-Queen's University Press, 2011); S. Max Edelson, *The New Map of Empire: How Britain Imagined America Before Independence* (Harvard University Press, 2017); Alexander Johnson, *The First Mapping of America: The General Survey of British North America* (I.B. Tauris, 2017); and James Lees, "Uprooting the Forest: Timber Surveying and State Formation in the British Atlantic, 1608–1806" (PhD diss., University of Oxford, 2024).

6 See, for example, John Stewart, *An Account of Prince Edward Island in the gulf of St. Lawrence* (London: W. Winchester & Son, 1806) and *Information to emigrants: an account of the*

island of Prince Edward, with practical advice to those intending to emigrate ... By Late resident of that colony (London: James Asperne, 1819).

7 "Capt. Holland's account of the Island of St. John," in Lockerby and Sobey, *Samuel Holland,* 173–74. The Island itself would be renamed from the Island of St. John's (the anglicized Île Saint-Jean) to Prince Edward Island in honour of the Duke of Kent, commander in chief of British forces in North America, who never visited the place. Somewhat ironically, given the enormous emphasis on freshwater river sources in these early sources, PEI today is totally dependent on groundwater for its freshwater supply.

Despite Holland's thorough renaming, a century later directories and tourist guides still referred to the rivers of Charlottetown by their more colloquial cardinal descriptors, as in this slightly confusing 1864 example describing the capitol as "looking toward the south[,] on the north of the East River, and near its junction with the North and West Rivers," *Hutchinson's Prince Edward Island Directory* (1864), 71.

8 On Wright's plan, see J.M. Bumsted, *Land, Settlement, and Politics on Eighteenth-Century Prince Edward Island* (McGill-Queen's University Press, 1987).

9 John Lawson, *Letters on Prince Edward Island* (Charlottetown, 1851), 62–63. Lawson admitted to his reader that "We want, it is true, a good many things, to bring us on a par with towns in your country. Gas, for one thing; this, however, we are told, we shall get for next to nothing, by Dr. Gesner's invention of Kerosene" – referring to Abraham Gesner's development of "coal oil" in neighbouring Nova Scotia. On "fruits to market," C. Birch Bagster, *The progress and prospects of Prince Edward Island, written during ... a visit in 1861* (Charlottetown: John Ings, 1861), 40. See also Edward MacDonald, "Charlottetown and Confederation," in *The Landscapes of Charlottetown,* ed. Catherine Hennessey, David L. Keenleyside, and Edward MacDonald (PEI Museum and Heritage Foundation, 2010); on Island traffic to Pictou and northern Nova Scotia as "coal country" in this era, Ruth McKenzie, ed., *The St. Lawrence Survey Journals of Captain Henry Wolsey Bayfield, 1829–1853* (Publications of the Champlain Society, 1984), and Daniel Samson, *The Spirit of Industry and Improvement: Liberal Government and Rural-Industrial Society, Nova Scotia, 1790–1867* (McGill-Queen's University Press, 2008).

10 Duncan Campbell, *History of Prince Edward Island* (Charlottetown: Bremner Bros., 1875), 206. For a laundry list of contemporary water-related concerns, see "Civic Requirements," Charlottetown *Examiner,* 24 May 1875, 2.

11 Lawson, *Letters on Prince Edward Island,* 29–30; letters to the editor, *Haszard's Gazette,* 23 May 1855, 2–3 and *Examiner,* 7 October 1867, 3; discussion of city council proceedings, *Examiner,* 11 January 1869, 3.

12 "The Spring Park Nuisance Case," *Examiner,* 27 October 1883, 2; "Spring Park Stream," *Examiner,* 21 October 1886, 2.

13 The waterworks was located on Hazard's Creek or Three Mile Brook, part of the Ellen Creek watershed. On the waterworks, see *Frederick's Prince Edward Island directory and book of useful information for 1889–90,* 51; now a municipal heritage site, the city describes the Charlottetown Water Works as "a beautiful landmark ... one of the most recognizable and unique heritage sites in Charlottetown" (which is saying something in a city that also boasts arguably the most recognizable heritage site in Canada, in Province House), https://www.historicplaces.ca/en/rep-reg/place-lieu.aspx?id=5641. For more on the development of waterworks and sewer systems in late nineteenth-century Charlottetown, see Douglas Baldwin, "The Charlottetown Political Elite: Control from Elsewhere" and "'But Not a Drop to Drink': The Struggle for Pure Water," in Baldwin and Spira, *Gaslights, Epidemics,*

and Vagabond Cows. In 1869, the town had granted licences to two men to serve "Citizens who may notify them of their desire to be supplied with pure water from the City Spring," albeit with some concern about privatizing water supply. Others – like Gilbert Murdoch, who produced the *Report on the Water Supply for the City of Charlottetown* in 1881 – felt that water sources further inland, further above urban and proto-urban contamination, were preferable.

14 *Haszard's Gazette,* 18 November 1851, 2.

15 Jim Hornby, *Black Islanders: Prince Edward Island's Historical Black Community* (Institute of Island Studies/University of Prince Edward Island, 1991). "Black Sam's Bridge" crossed over Spring Park at Euston Street/Brighton Street at the head of Government Pond.

16 Grumbler [pseud.], "A Growl," *Examiner*, 24 January 1882, 2; *Examiner*, 16 November 1886, 3; "Plinthos," "Victoria Park Drive," *Examiner*, 23 April 1887, 1; "Observer," *Examiner,* 31 January 1878, 2; "The Spring Park Nuisance Case," *Examiner,* 27 October 1883, 2.

17 "Improvements," *Examiner,* 11 June 1898, 4. On eighteenth- and nineteenth-century attitudes toward wetlands, see Rodney J. Giblett, *Postmodern Wetlands: Culture, History, Ecology* (Edinburgh University Press, 1996); Ann Vileisis, *Discovering the Unknown Landscape: A History of America's Wetlands* (Island Press, 1997); Melanie Kiechle, *Smell Detectives: An Olfactory History of Nineteenth-Century Urban America* (University of Washington Press, 2017); and Eric H. Ash, *The Draining of the Fens: Projectors, Popular Politics, and State Building in Early Modern England* (Johns Hopkins University Press, 2017).

18 "The Park for Bathers," *Examiner*, 5 August 1882, 2.

19 "I have been told," the writer added, not being above flattery, "that this water side of our town is very like one of the views in the beautiful harbor of Copenhagen." "The City Park," *Examiner,* 29 May 1876, 2. Harry Holman and Edward MacDonald suggest that "W.C.H." is William Critchlow Harris, whose family was already investing in properties in the new Brighton suburb west of the pond.

20 Harris had, for example, designed "Beaconsfield" for shipping magnate James Peake in 1877 at the end of Kent Street, where it reached the mouth of the pond. "Government Pond Again," *Examiner*, 6 December 1886, 1. His comparatively nuanced, progressive recommendation included a drain running from the high-tide mark along the entire length of the brook to Malpeque Road, to collect "all foul water from the district" and allow spring water to flow unimpeded to the shore; carting out the pond mud as fertilizer; and planting willows around the water's edge. The resulting avenue would make Charlottetown "one of the most beautiful cities in Canada." Though he did not reference Copenhagen, Harris pointed to Halifax's design of the Public Gardens – with a pond artfully made from another stream (Freshwater Brook) and avenues lined with willows.

21 As early as 1886, Harris recalled how "the pond above the Brighton Bridge, between the two tanneries ... was at one time covered with the ebb and flow of salt water." This points to both the extent of estuarine and tidal flow and the long-standing campaign to cover the wet surfaces.

22 "The New Bridge," *Examiner,* 8 September 1900, 4; "Spring Park Brook in Flood," *Guardian,* 25 March 1949, 3.

23 Some of the material in this part of the chapter previously appeared in an article I wrote: "Hardened Water: The Remaking of a Coastal City," *Coastal Studies & Society* 2, 1 (2022): 112–36. My thanks to the journal.

24 On the gasworks, "City Gasworks – Recent Improvements – Buildings and Apparatus Renewed," *Examiner*, 20 February 1880, 2. On cattle pasturage, Acadians and early British

settlers had pastured their cattle extensively on salt marsh hay, though by the mid-nineteenth century tidal flats were more likely to be scraped for their rich mud, which was used as fertilizer to rejuvenate the upland soils worked ever more intensely for growing potatoes and other crops. Still, it appears the older tradition of pasturage was retained on the city's outskirts. See the Meacham Atlas of 1880 and the Cummins Atlas of 1929; "Falconwood," Historic Places Register, https://www.historicplaces.ca/en/rep-reg/place-lieu.aspx?id=7481; Wayne MacKinnon and Elinor Vass, "The Best of the Past: Traditional Sustainable Agriculture in Prince Edward Island" (Institute of Island Studies, 1989). For more on the changing agricultural demands on salt marshes and tidal flats, see Matthew Hatvany, "'Wedded to the Marshes': Salt Marshes and Economic Differentiation in Early Prince Edward Island" *Acadiensis* 30 (2001): 40–55; Rusty Bittermann and Margaret McCallum, "'One of the Finest Grass Countries I Have Met With': Prince Edward Island's Colonial-Era Cattle Trade," *Agricultural History* 90, 2 (2016): 173–94; and Joshua MacFadyen, "Drawing Lines in the Ice: Regulating Mussel Mud Digging in the Southern Gulf of St. Lawrence," in *Land & Sea: Environmental History in Atlantic Canada*, ed. Claire Campbell and Robert Summerby Murray (Acadiensis Press, 2013). An abattoir and, later, meat-packing plant sat on this location until late in the twentieth century.

25 Rusty Bittermann and Margaret E. McCallum, "The One That Got Away: Fishery Reserves in Prince Edward Island," *Dalhousie Law Journal* 28, 2 (2005): 389.

26 "The New Railway Station," *Guardian*, 9 April 1904, 2; "Our Civic Rulers," *Examiner*, 22 May 1888, 3; also *Examiner*, 6 May 1880, 3; "A Nuisance," *Examiner*, 19 May 1885, 3 and "A Nuisance," *Examiner*, 3 May 1889, 3.

27 "Sewage Wells," *Examiner*, 25 November 1889, 2.

28 "City Improvements," *Examiner*, 2 October 1876, 2; "Our Driving and Exhibition Park," *Examiner*, 17 January 1889, 2. The marsh, though generally unnamed, was occasionally referred to as Beers' Marsh or Beasley Marsh.

29 For example, there were regularly items in the annual budget for fill, breastwork, or "filling up space covered by water in the station yard," as in the Minister of Railways and Canals 1887/01/24 Privy Council Minutes, 25 Janaury–1 February 1887, Library and Archives Canada.

30 Canada House of Commons, *An Act to Authorize the Construction of a Branch Railway from Charlottetown to Murray Harbour, as a Public Work*, 1899; reprinted "Murray Harbour Branch, Hill-borough [sic] Bridge at Charlottetown, P.E.I.," report from W.B. MacKenzie, Chief Engineer, Moncton, to D. Pottinger, General Manager, Government Railways, *Examiner*, 27 February 1900, 5; "The Bridge," *Examiner*, 16 March 1900, 4 and 5. See Rosemary Curley, "Expansion of Charlottetown Along the Hillsborough Shore," *Island Magazine* 58 (2005): 28–31; J.P. McCloskey, "Bridging the Hillsborough," *Island Magazine* 56 (2004): 32–37; and "Construction of the Prince Edward Island Railway," Canada Historic Sites and Monuments Board, 2018, https://www.pc.gc.ca/en/culture/clmhc-hsmbc/res/information-backgrounder/rail_IPE-PEI-railway. As at Portland, Maine, the arrival of the railway on the city's outskirts would dramatically alter the spatial and economic dynamics of the city. See Martyn J. Bowden, "Industrial Portland," Plate 45, in Stephen J. Hornsby and Richard W. Judd, eds., *Historical Atlas of Maine* (University of Maine Press, 2015).

31 "The Station Site," *Guardian*, 20 December 1904, 4.

32 Hansard, 5 July 1904, House of Commons *Debates*, 9th Parliament, 4th Session, 5997.

33 The federal government had purchased the land – or pond, or whatever – between the railway and the shore of the province. Hansard, 3 March 1905, House of Commons *Debates*, 10th Parliament, 1st Session, vol. 2, 2025.

34 Subject to civic regulations; so, presumably, no bodies. *Examiner,* 23 June 1884, 3; "Prince Edward Island Railway," *Guardian,* 5 March 1904, 5; "PEI Railway: Tender for Filling," *Impartial* and *Guardian,* October–November 1905.

35 News items, *Guardian,* 24 June 1905, 9; 8 July 1905, 5; 25 July 1905, 5. Also Prince Edward Island Railway – acceptance of tender of J.B. McManus L't'd for the filling of a pond east of the Station at Charlottetown on – Min. R. and C. [Minister of Railways and Canals], 1906/05/15 (RG2, Privy Council Office, Series A-1-a), Library and Archives Canada. On the accidents, see "Untimely End of Brilliant Young Man," *Guardian,* 7 December 1904, 1; "Nearly a Fatality," *Guardian,* 4 April 1919, 7.

36 David Harvey, *The Limits to Capital* (Verso, 2006), 233.

37 Annual Reports of the Department of Railways and Canals, 1899–1900 through 1910–11, Sessional Papers; "New Station to Be Opened Today," *Guardian,* 8 July 1907, 1–3. On reading insurance maps, see G.T. Bloomfield, "Canadian Fire Insurance Plans and Industrial Archeology," *IA: The Journal of the Society for Industrial Archeology* 8, 1 (1982): 67–80; Bob Hayward, "Chas. E. Goad and Fire Insurance Cartography," in *Explorations in the History of Canadian Mapping: A Collection of Essays,* ed. Barbara Farell and Aileen Desbarats (Association of Canadian Map Libraries and Archives, 1988); John Boylan, "The Best Laid Plans: Fire Insurance Mapping on Prince Edward Island," *Island Magazine* 58 (2005): 23–27; and Jason Gilliland and Mathew Novak, "Positioning the Past with the Present: The Use of Fire Insurance Plans and GIS for Urban Environmental History," *Environmental History* 11, 1 (2006): 136–39.

38 Claire Campbell, "Whatever Happened to Pleasant Street? Rediscovering an Urban Shoreline," *Environmental History* 25, 1 (2020): 134–49. See chapter 4.

39 Halifax Municipal and Regional Archives, CR30D.115 Heritage Trust Brief re. Water St. Wharf Buildings, Civic Advisory Committee, November 1962.

40 "Introduction: Resilience, Disaster, and Rebuilding in Modern Port Cities," in "Resilience in Seaports," ed. Carola Hein and Dirk Schubert, special issue, *Journal of Urban History* 47, 2 (2021): 235–49. This aligns with David Harvey's idea of "fixing" or securing capital in place in order to help stabilize and sustain capitalism as a system. Harvey, *Limits to Capital.*

41 The power plant was not decommissioned until 2017. *Environmental Impact Statement for Concert Venue Site and Parking Facility, Imperial Oil Co. and Adjacent Lands* (Charlottetown Area Development Corporation, 2010).

42 These include the PEIR brass house (1876), car shop (1906), and the railway station (1907). See the Canadian Register of Historic Places (www.historicplaces.ca) and Charlottetown Area Development Corporation (http://cadcpei.com/projects/cn-car-shop-founders-hall/). David Harvey assesses another port's gentrification – albeit one with a very different racial and political history – at Baltimore in "A View from Federal Hill," first published in *The Baltimore Book: New Views of Local History,* ed. Elizabeth Fee, Linda Shopes, and Linda Zeidman (Temple University Press, 1991), 227–52.

43 Stephanie Kane describes container ships as "colossal floating commodity dens," but the same could be said of the cruise ship. Kane, *Where Rivers Meet the Sea: The Political Ecology of Water* (Temple University Press, 2012), 8. See also Alan MacEachern and George Edward MacDonald, *The Summer Trade: A History of Tourism on Prince Edward Island* (McGill-Queen's University Press, 2022).

44 See https://www.urbanstrategies.com/project/eastern-gateway-waterfront-master-plan/. Because infill can be more unstable than "natural" bedrock or sedimentary ground, parking is arguably a safer choice than building weight.

45 See "Low Flood Hazard [2100]," in PEI Climate Hazard and Risk Information System [CHRIS], https://chris.peiclimate.ca.

46 Michael J. Chiarappa and Kristin M. Szylvian, "Heeding the Landscape's Usable Past: Public History in the Service of a Working Waterfront," *Buildings & Landscapes: Journal of the Vernacular Architecture Forum* 16, 2 (2009): 86–101; and R. Timothy Sieber, "Waterfront Revitalization in Postindustrial Port Cities of North America," *City & Society* 5, 2 (1991): 120–36.

47 In 2021, an "active transportation lane" was added to the Hillsborough Bridge.

48 Andrew Hurley, *Beyond Preservation: Using Public History to Revitalize Inner Cities* (Temple University Press, 2010), 127, 144.

Chapter 4: Tidewater Aligned

1 Marc Levinson, *The Box: How the Shipping Container Made the World Smaller and the World Economy Bigger,* 2nd ed. (Princeton University Press, 2016).

2 Some of the ideas in this chapter were developed in Claire Campbell, "Whatever Happened to Pleasant Street? Rediscovering an Urban Shoreline in Halifax, Nova Scotia," *Environmental History* 25, 1 (January 2020): 134–49; my thanks to the journal. There are several claims to the first and fastest steam-powered Atlantic crossing; Cunard was also part-owner of the *Royal William,* which sailed from Pictou and arrived at Gravesend after twenty-five days in August–September 1833.

3 See, for example, *The Writer's Map: An Atlas of Imaginary Lands*, ed. Huw Lewis-Jones (University of Chicago Press, 2018).

4 Ian McKay and Robin Bates, *In the Province of History: The Making of the Public Past in Twentieth-Century Nova Scotia* (McGill-Queen's University Press, 2010), 201. Raddall was best known for his historical fiction based on the age of sail and the wars of the eighteenth century; see Alan R. Young, ed., *Time and Place: The Life and Works of Thomas H. Raddall* (Acadiensis Press, 1991). For other literary excerpts set in Halifax, see John Bell, ed., *Halifax: A Literary Portrait* (Pottersfield Press, 1990).

5 For example, see David Sutherland, "Halifax 1815–1914: 'Colony to Colony,'" *Urban History Review/Revue d'histoire Urbaine* 4, 1 (1975): 7–11.

6 Judith Fingard, Janet Guildford, and David Sutherland, *Halifax: The First 250 Years* (Formac Publishing, 1999). See note 74 for the Parks Canada research. Archaeologist Jonathan Fowler leads a number of public-facing projects in downtown Halifax as well.

Halifax is in the background of Mark Leeming's *In Defence of Home Places: Environmental Activism in Nova Scotia* (UBC Press, 2017) as home to the Ecology Action Centre, though the issues of concern are not themselves urban. The exception would be Africville, which has been treated by scholars of race and urban planning as well as historians, though primarily focused in the postwar era.

7 There is also a lesser-trafficked path along the Dartmouth waterfront, part of the TransCanada Trail.

8 Margaret Conrad, *At the Ocean's Edge: A History of Nova Scotia to Confederation* (University of Toronto Press, 2020). Named for the Earl of Halifax, who at the time was president of the Board of Trade and Plantations.

9 Thomas Haliburton, *A general description of Nova Scotia* (Halifax: Clement Belcher, 1825), 60.

10 Julian Gwyn, *Ashore and Afloat: The British Navy and the Halifax Naval Yard Before 1820* (University of Ottawa Press, 2004), 7–16; Marilyn G. Smith, *The King's Yard: An Illustrated History of the Halifax Dockyard* (Nimbus Publishing, 1985); and D.R. Moore, "History of HMC Dockyard" (n.p., ca. 1970), 971.6225.M, Halifax Municipal Archives (HMA). On the context of the eighteenth century and the new military landscapes of Acadia, see Thomas Peace, *The Slow Rush of Colonization: Spaces of Power in the Maritime Peninsula, 1680–1790* (UBC Press, 2023).

11 Hood to Navy Board, 13 October 1767, cited in Gwyn, *Ashore and Afloat,* 13. See, by way of illustration, the "Plan of the environs of Halifax in Nova Scotia with the temporary works constructed for the protection of the town, and His Majesty's naval yard" (1779), William L. Clements Library Image Bank, University of Michigan.

12 Gwyn, *Ashore and Afloat,* 225.

13 David Jaffee, *People of the Wachusett: Greater New England in History and Memory, 1630–1860* (Cornell University Press, 1999), 179–80.

14 John Wentworth to Crown Land Grant Old Book 29 p 9, 23 September 1794.

15 *A map of the south part of Nova Scotia and its fishing banks engraved by T. Jefferys geographer to His Royal Highness the Prince of Wales,* 1750, detail, John Carter Brown Library. Standing water is visible in two of the insets: "A Plan of Halifax Survey'd by M.[oses] Harris" and "A View of Halifax Drawn from Ye Topmasthead." See also Susan Buggey, "The Restoration Site," *Halifax Waterfront Buildings,* Occasional Papers in Archaeology and History no. 9 (National Historic Parks and Sites Branch, Parks Canada, 1974).

16 Fingard et al., *Halifax: The First 250 Years,* 50.

17 Thomas B. Akins, *History of Halifax City* (Halifax, 1895), 201. Akins notes that the road past Steele's Pond had at one point been washed out by the tide and an older road may have followed the shore even more directly until the south escarpment was occupied by the Royal Navy's lumberyard (202, 213). The south end of the peninsula was formally leased to the city by the Crown in 1866 and became known as Point Pleasant Park.

18 Halifax incorporated as a city in 1840 with a population of about fifteen thousand. Susan Buggey, "Building Halifax, 1841–1871," *Acadiensis* 10, 1 (1980): 112.

19 Sherry Olson, "Downwind, Downstream, Downtown: The Environmental Legacy in Baltimore and Montreal," *Environmental History* 12, 4 (2007): 850.

20 Between 1861 and 1871, Halifax vied with Saint John for the claim as the fourth largest city in British North America after Montreal, Quebec, and Toronto. "Table IV: Population of Cities and Towns having over 5,000 inhabitants compared," Census of Canada, 1871.

21 "The Petition of the People of the County of King's," included in the *Letter to the Earl of Carnarvon by Mr. Joseph Howe, Mr. William Annand, and Mr. Hugh McDonald Stating their Objections to the Proposed Scheme of Union of the British North American provinces, Presented to both Houses of Parliament,* 8 February 1867 (London: Eyre and Spottiswoode, 1867), 25; Del Muise, "The Federal Election of 1867 in Nova Scotia: An Economic Interpretation," *Collections of the Royal Nova Scotia Historical Society* 36 (1968): 327–29; and Philip Buckner, "The Maritimes and Confederation: A Reassessment," *Canadian Historical Review* 71, 1 (1990): 1–30. Although ratified by a vote in the colonial legislature (i.e., not a popular vote) in 1866, the depth of public opposition to the terms of Confederation became clear when Nova Scotia held its first popular elections the next year, less than three months after the British North America Act had come into effect. Joseph Howe and the anti-Confederates swept the provincial legislature, winning thirty-six of thirty-eight seats, and

Nova Scotia filled eighteen of its nineteen allocated seats in the new Dominion Parliament with anti-Confederate Liberals. For the next fifty years, especially during economic downturns, provincial elections were frequently won on the promise to campaign for "better terms" (usually larger subsidies) or even outright secession.

22 Hugh MacLennan, *Barometer Rising*, New Canadian Library ed. (McClelland & Stewart, 1989), 7. Sail and steam ran parallel for a few decades, while sail gradually became reserved for slower and bulk cargo. By the 1870s, in the bigger port cities such as Liverpool and New York it had lost its competitive edge to steam on both water and the new rail connections on land. Melissa Meriam Bullard, *Brooklyn's Renaissance: Commerce, Culture, and Community in the Nineteenth-Century Atlantic World* (Palgrave Macmillan, 2017), 67–70.

23 The dockyard had had a careening yard – able to tilt ships to work on their hulls – since the 1750s. See E.H. Keating, City Engineer, "Preliminary Report on the Halifax Dry Dock and Report on American Stone and Wooden Docks" (Halifax: Nova Scotia Printing, 1883); Tom Tulloch, "The Halifax Graving Dock and the 1917 Explosion," *Argonauta* (Canadian Nautical Research Society) 36, 1 (2019): 6–22; and Steven Gray, *Steam Power and Sea Power: Coal, the Royal Navy, and the British Empire, 1870–1914* (Palgrave Macmillan, 2018). For more on the energy history of Nova Scotia, see Claire Campbell, "Privileges and Entanglements: Lessons from History for Nova Scotia's Politics of Energy," *Acadiensis* 42, 2 (2013): 114–37.

24 William B. Meyer, "A City (Only Partly) on a Hill: Terrain and Land Use in Pre-Twentieth Century Boston," in *Remaking Boston: An Environmental History of the City and Its Surroundings*, ed. Anthony Penna and Conrad Wright (University of Pittsburgh Press, 2009).

25 *Halifax - Nova Scotia: The Garrison City by the Sea* (Nova Scotia Tourist Association, 1900), 30 (emphasis added).

26 *A Guide to the city of Halifax, Nova Scotia, 1884 & 1885: Its past and present history* (Halifax: William McNab, 1884), 28 (emphasis added). *The Garrison City by the Sea* also notes a beach at the foot of Salter Street, in the downtown core, that eighteenth-century maps show as the outlet of an unnamed stream. The lot at the mouth of Freshwater was sold to the city in 1860 by Alderman Charles Cogswell, who also sponsored rowing races on the harbour.

City Council Minutes, "Resolution relative to paying Dr. Cogswell for Water Lot at Fresh Water," 1860-07-03, page 114. The sea wall was built by 1870 (*Halifax Daily Citizen*, 23 November 1870, 3). The esplanade and the mouth of Freshwater Brook appear to have survived into the twentieth century as a city park (sandwiched between a planing mill and a distillery), as a tacit concession to regular flooding in the Freshwater corridor. Sheet 27, Charles E. Goad, Insurance Plan, City of Halifax, Nova Scotia (May 1895, revised to March 1899), Nova Scotia Archives (NSARM). This intersection of Barrington and Inglis Streets – directly over the original outlet of Freshwater Brook – regularly floods during rain events.

27 Judith Fingard, *The Dark Side of Life in Victorian Halifax* (Pottersfield Press, 1989); Stanford Eveleth, *Miss Dexie: A Romance of the Provinces* (W. Briggs, 1895); and Gwendolyn Davies, "Emma Lucy Wells (Dickson)," *Dictionary of Canadian Biography*, vol. 15 (University of Toronto/Université Laval, 2003–).

28 Lucy Maud Montgomery, *Anne of Green Gables* (L.C. Page, 1908).

29 Lucy Maud Montgomery, *A Name for Herself: Selected Writings, 1891–1917*, ed. Benjamin Lefebvre (University of Toronto Press, 2018).

30 Kate Scarth, "Born for City Life? L.M. Montgomery's Urban Prince Edward Island," *Island Magazine* 86 (2020):50–60. For more on Montgomery's depiction of nature, especially the shoreline, see Claire Campbell, "'A Window Looking Seaward': Finding Environmental

History in the Writing of L.M. Montgomery," in *The Greater Gulf: Essays on the Environmental History of the Gulf of St. Lawrence*, ed. Claire Campbell, Edward MacDonald, and Brian Payne (McGill-Queen's University Press, 2020), 283–318.

31 "After the Ball," *Halifax Daily Echo*, 26 October 1901, 1, 12; see Montgomery, *A Name for Herself*, 89–93.

32 "A Walk in the Park," *Halifax Daily Echo*, 23 November 1901, 1; see Montgomery, *A Name for Herself*, 103–6.

33 *American Young People* (1903?), republished in *Western Christian Advocate* (11 May 1910) and in Rea Wilmshurst, ed., *At the Altar: Matrimonial Tales* (McClelland & Stewart, 1995), 105–14. My thanks to Simon Lloyd of Robertson Library, UPEI, for the reference.

34 Lucy Maud Montgomery, *Anne of the Island* (A.L. Burt, 1915).

35 Montgomery, *Anne of the Island*, 30.

36 Aaron Sachs, *Arcadian America: The Death and Life of an Environmental Tradition* (Yale University Press, 2013); and Ron Williams, *Landscape Architecture in Canada* (McGill-Queen's University Press, 2014).

37 Montgomery, *Anne of the Island*, 31, 214, 63.

38 Montgomery, *Anne of the Island*, 62. It also allowed her to draw verbatim – as she often did – from her diaries. See the entry for 23 December 1895 in Lucy Maud Montgomery, *The Complete Journals of L.M. Montgomery, the PEI Years, 1889–1900*, ed. Mary Rubio and Elizabeth Waterston (Oxford University Press, 2017), 298–99.

39 The replica *Bluenose II* (built 1963) and MacAskill photographs remain prominent and widely circulated symbols of Nova Scotia today. F.W. Wallace, *Wooden Ships and Iron Men: The Story of the Square-Rigged Merchant Marine of British North America, the Ships, Their Builders and Owners, and the Men Who Sailed Them* (Hodder and Stoughton, 1924); Ian McKay, *The Quest of the Folk: Antimodernism and Cultural Selection in Twentieth Century Nova Scotia* (McGill-Queen's University Press, 1994); and M. Brook Taylor, *A Camera on the Banks: Frederick William Wallace and the Fishermen of the Grand Banks* (Goose Lane Editions, 2000).

40 Anne's friend Philippa marries a young minister who intends to take up a "little mission church" in "the slums" (231) north of the Citadel, but we only hear briefly about this part of the city.

41 Thomas Raddall, *Halifax: Warden of the North* (Doubleday, 1965), 235. The Royal Navy organized the city's first formal regatta in 1826, with the Halifax Yacht Club organized a decade later (and renamed the Royal Nova Scotian Yacht Squadron after the Prince of Wales' visit in 1860). See *A Map of the City and Harbour of Halifax* (C.D. McAlpine, 1910). HRMA 102-5A-26.

42 Caitlin Charman, "'An ugly, piled-up sea': Industrialization and Regional Identity in W. Albert Hickman's Gulf of St. Lawrence Fiction," in Campbell et al., *Greater Gulf*. *Powering Up: A History of Fuel, Power and Energy in Canada, 1800–2015*, ed. Ruth Sandwell (McGill-Queen's University Press, 2016), the first collection on energy history in Canada, reviews both organic and mineral regimes. Like most other resource sectors in Canada, the study of energy (especially coal) has tended to be framed as sites of constitutional competition or labour mobilization.

43 "Some Facts About Nova Scotia's Oil Field," *Halifax Herald*, 25 April 1903, 5.

44 "Industrial Supplement," *Acadian Recorder*, August 1902, 16.

45 *McAlpine Halifax City Directory* (Halifax, 1916), NSARM; "Oil Is King for Democracy to Rigorously Rule," *The Evening Mail*, 15 December 1915, 9. Exploratory drilling for petroleum

also began in Cape Breton and on the Bay of Fundy. "Maritime Oil & Gas Co. Ltd.," *Evening Mail*, 1 January 1910, 9. There was already interest in generating electricity from the tidal currents in the Bay of Fundy, home to the highest tides in the world; see "Power from Tidal Currents in Bay of Fundy," *Journal of the Royal Society of Arts* 64, 3333 (1916): 788. Over a century later, tidal power has remained elusive because the tides are too strong for the turbines.

46 Del A. Muise, "'The Great Transformation': Changing the Urban Face of Nova Scotia, 1871–1921," *Nova Scotia Historical Review* 11, 2 (1991): 1–42.

47 See, for example, John S. Dryzek and Jonathan Pickering, *The Politics of the Anthropocene* (Oxford University Press, 2018); Sebastian Haumann, Martin Knoll, and Detlev Mares, eds., *Concepts of Urban-Environmental History* (Transcript, 2020); and "Introduction: Resilience, Disaster, and Rebuilding in Modern Port Cities," in "Resilience in Seaports," ed. Carola Hein and Dirk Schubert, special issue, *Journal of Urban History* 47, 2 (2021): 235–49.

48 Mark A. Greene, *The Great Saint John Fire of 1877: The Rise, Destruction and Recovery of Canada's Leading Port City* (Formac Publishing, 2022); also Chapter 2 for fires at St. John's. The "Great Fire" of Toronto in 1904 would also have been well known in Halifax.

49 MacLennan, "An Orange from Portugal," originally published 1978; in *An Orange from Portugal: Christmas Stories from the Maritimes and Newfoundland*, ed. Anne Simpson (Goose Lane Editions, 2003), 223.

50 "Industrial Supplement," *Acadian Recorder*, 2.

51 *Halifax Nova Scotia and Its Attractions* (Howard and Kutsche, 1904), n.p.

52 See Steven Schwinghamer and Jan Raska, *Pier 21: A History* (University of Ottawa Press, 2020). In 1903, as part of a long and convoluted debate over a proposed third national railway, Robert Borden, then Member for Halifax and leader of the Conservative Opposition, told the House of Commons, "I know of no better way in which the money of this country can be spent than by improving the great inland waterway, the St. Lawrence route, and our great national ports." House of Commons *Debates* 18 August 1903, 9004. New York had experienced the same anxiety about freight capacity and wharf space. As Ann Buttenwieser writes, "There was a constant entreaty to push the edges of the island both seaward and skyward." *Manhattan Water-Bound: Manhattan's Waterfront from the Seventeenth Century to the Present*, 2nd ed. (Syracuse University Press, 1999), 73. Ted Steinberg phrases this as "destiny demanded destiny," in *Gotham Unbound: The Ecological History of Greater New York* (Simon & Schuster, 2015), 184.

53 A "General Features Terminal Improvements" plan from 1910 proposed extending twin lines into the downtown core to divorce and expand freight and passenger service while "Centraliz[ing] both Services in the heart of the City at water front level." Map Collection V6 240, NSARM.

54 "Transportation Routes in Canada," *The Journal of the Engineering Institute of Canada* 11, 7 (July 1923): 313–22. Also *Who's Who and Why: A Biographical Dictionary of Men and Women of Canada and Newfoundland*, vol. 6–7, *1915–16*, ed. C.W. Parker (International Press, n.d.), 353; Cowie, "Report on the River St. Lawrence Ship Channel from Montreal to Quebec and the Traverse" (Ottawa, 1906). Cowie would later draft a proposal for the port of Savannah.

55 *Halifax Herald*, 31 October 1912, 1–2; "The New Order in Halifax Will Be Business Before Pleasure," *Herald*, 4 November 1912, 2. The *Herald* certainly felt that "Hon. Messrs. Borden [Prime Minister Robert Borden] and Cochrane [Frank Cochrane, Minister of Railways and Canals] are the Best Friends Nova Scotia Ever Was Blest With" (31 October 1912, 6).

56 "The Terminals Proposed by Government for Halifax Ensure Our Future," *Halifax Herald*, 19 November 1912, 3; "Our Terminals Unexcelled on Atlantic Seaboard," *Evening Mail*, 16 June 1914; Barry Cahill, "Mackintosh, James Crosskill," in *Dictionary of Canadian Biography*, vol. 15 (University of Toronto/Université Laval, 2003–). See also Sandford Fleming, letter to the city/mayor of Halifax, read into the Council minutes (18 November 1912), HMA. The Liberal-leaning papers were unconvinced, although they expressed their criticism in similarly extravagant language: the *Morning Chronicle* called the choice of site "a colossal blunder" (8 January 1913, 3) and the *Nova Scotian* insisted that the old Pier 2 at the Deep Water terminals in the Narrows was still "One of the Best in the World" (5 September 1912, 2).

57 Frederick William Cowie, *Report to the Honourable Frank Cochrane, Minister of Railways and Canals, on Halifax Harbour and* (Montreal, 1913), 8, 47. The most thorough history of the port and its aspirations to be Canada's winter port and eastern gateway – commissioned by the Halifax Port Authority and with no reference to environmental impact – is James D. Frost's understandably celebratory *Canada's Atlantic Gateway: An Illustrated History of the Port of Halifax* (Nimbus Publishing, 2008).

58 Cowie, *Report to the Honourable Frank Cochrane*, 11, 49. The idea of "fixed capital" also recalls David Harvey's arguments about how capitalism protects itself through investments in infrastructure that further circulate capital, reach new markets, etc. David Harvey, *The Limits to Capital* (Verso, 2006).

59 On the industrial sublime, see David E. Nye's *American Technological Sublime* (MIT Press, 1994). It is also a term frequently associated with the Canadian photographer Edward Burtynsky; see, for example, the catalogue for an exhibition of the same name (Weber State University, 2011). Burtynsky's choice of an aerial perspective to depict the petroscape both seems necessary to understand the scale of the industrial project and has the effect of distancing us further from the ground.

60 Carrick Limited, "Halifax: An Old City with a Brand New Future" (1913), Halifax Public Libraries. In the process of constructing the terminals, some houses were built in a neighbourhood above the piers that became known as "Greenbank." These were razed by the 1960s when the land was acquired by the Halifax Port Corporation to expand southward from the terminals project for more container space.

61 "Well and Truly Laid by Premier Borden," *Evening Mail*, 21 October 1915, 5 (emphasis added); also "The Mammoth Ocean Terminals Which Crown the Harbor of Halifax," 16 May 1917, 8.

62 MacLennan, *Barometer Rising*, 196.

63 See "Contaminants Found in Testing on Halifax Common Could Date Back to Explosion," *CBC News*, 20 March 2021, https://www.cbc.ca/news/canada/nova-scotia/contamination-halifax-common-1.5954476; Strum Consulting, "Environmental Screening Report, Halifax Aquatic Centre, Cogswell Street, Halifax, NS (PID 00137489)," 2020, obtained from Halifax Regional Municipality.

On the Explosion, one might begin with David Sutherland, ed., "*We Harbor No Evil Design": Rehabilitation Efforts After the Halifax Explosion of 1917* (The Champlain Society, University of Toronto Press, 2017), 12. Sutherland's introduction is an excellent account of the neighbourhood geographies of Halifax before the explosion as well as the events of 1917 and following. See also Jacob A.C. Remes, *Disaster Citizenship: Survivors, Solidarity, and Power in the Progressive Era* (University of Illinois Press, 2016).

The scholarly literature is dwarfed by the popular history inspired by the event. Consider even in the past twenty years: Laura MacDonald, *The Curse of the Narrows: The Halifax*

Explosion, 1917 (2005); Janet Kitz, *Shattered City: The Halifax Explosion and the Road to Recovery* (2010); Michael Bird, *The Town That Died: A Chronicle of the Halifax Explosion* (2011); Ken Cuthbertson, *The Halifax Explosion: Canada's Worst Disaster* (2017); John Boileau, *6/12/17: The Halifax Explosion* (2017); and the hybrid academic/journalistic *Catastrophe: Stories and Lessons from the Halifax Explosion,* by Joseph Scanlon with Roger Sarty (Wilfrid Laurier University Press, 2020). There are also accounts specifically from journalists, druggists, the Royal Canadian Navy, etc., in addition to children's stories and fiction set during the event.

64 The first person Raddall meets after staggering to his feet after the blast is an Imperial Oil delivery wagon. The wagon driver manages to rouse the stunned horses and they set off again, "leaving a small puddle of kerosene on the road." Thomas Raddall, *In My Time: A Memoir* (McClelland & Stewart, 1976), 33–35. Raddall elaborates on the mushroom cloud in "A Winter's Tale": "it extended an enormous distance upward from invisible roots in the harbour; and at the top it was unfolding, spreading out rapidly in greasy curls, brown and black, that caught the December sun and gleamed with a strange effect of varnish. An evil mushroom that writhed slightly on its stalk, and spread its eddying top until it overshadowed the whole North End, strange and terrible and beautiful." In *At the Tide's Turn and Other Stories* (McClelland & Stewart, 1971), 161–78.

65 "A Winter's Tale," 169.

66 Elspeth Cameron, *Hugh MacLennan: A Writer's Life* (University of Toronto Press, 1981); on the colony-to-nation reading, see, as an example, Robert MacNeil, "Creating Fiction Out of Fact," in *Ground Zero: A Reassessment of the 1917 Explosion in Halifax Harbour*, ed. Alan Ruffman and Colin D. Howell (Nimbus Publishing/Gorsebrook Research Institute for Atlantic Canada Studies, 1994), 124.

67 MacLellan, *Barometer Rising,* 40.

68 MacLellan, *Barometer Rising,* 12, 65, 184.

69 MacLellan, *Barometer Rising,* 190; Frederick McKelvey Bell, *A Romance of the Halifax Disaster* (The Royal, 1918), n.p.

70 MacLellan, *Barometer Rising,* 253.

71 *Le Port d'Halifax/Halifax Harbour,* ed. Anabelle Poñka (Goose Lane/National Gallery of Canada, 2018). As a correlation, a character in Howard Norman's 1998 novel *The Museum Guard* (Picador), set in 1930s Halifax, suggests Halifax is *not* really a city: "'All these trees. They make me want to get on a train to Toronto,' he said. 'I much prefer cities. I like when paintings of city life come in'" (34). This novel is framed by a fictional disaster contemporaneous with the explosion: a zeppelin crash in 1916 at the city fairgrounds.

72 Mary Elizabeth Luka, "Walking Matters: A Peripatetic Rethinking of Energy Culture," in *Energy Culture: Art and Theory on Oil and Beyond,* ed. Imre Szeman and Jeff Diamanti (West Virginia University Press, 2019), 65–66.

73 MacLellan, *Barometer Rising,* 159.

74 MacLellan, *Barometer Rising,* 111.

75 MacLellan, *Barometer Rising,* 75.

76 The Board of Trade touted all three of these – the dry dock, terminals under construction, and the refinery at Imperoyal – as tourist attractions for those automobiling about. Bordering on the inhumanely opportunistic, it suggested the site of the explosion offered "an exceptionally fine view of the harbor and basin." Halifax Board of Trade, *Halifax: The City by the Sea* (ca. 1920), 10–15.

77 The first Acadia refinery, originally located in the Narrows, had been obliterated by the explosion in 1917.

78 Mark Stoll, *Profit: An Environmental History* (Polity Press, 2023), 151.

79 An older distribution of lubricant and illuminating oils was quickly supplanted by petroleum and its byproducts. "The Tremendous Works on Halifax Harbour," *Halifax Herald*, 31 December 1917, 23. Graham D. Taylor, *Imperial Standard: Imperial Oil, Exxon, and the Canadian Oil Industry from 1880* (University of Calgary Press, 2019) makes almost no mention of the Dartmouth refinery but does provide a good corporate overview of the sector. Halifax was the first port in Canada to receive a cargo of oil by tank steamer: "A Record Oil Cargo," *Halifax Herald* (3 December 1901), 3. For a creative postindustrial reading of this oil landscape, see Tyler Reynolds, "Re-imagining a Petrochemical Landscape in Dartmouth Nova Scotia" (MArch thesis, Dalhousie University, 2016).

80 Raddall, *The Nymph and the Lamp* (Little, Brown, 1950), 8–9. Raddall had also served as a telegraph operator on ships making runs for coal from Newfoundland to Cape Breton to Manchester.

81 Christopher J. Armstrong, "Post-Colonial Historicity: Halifax, Region, and Empire in *Barometer Rising* and *The Nymph and the Lamp*," in *Downtown Canada: Writing Canadian Cities*, ed. Justin D. Edwards and Douglas Ivison (University of Toronto Press, 2005).

82 Raddall, *The Nymph and the Lamp*, 38.

83 Raddall, *The Nymph and the Lamp*, 326.

84 Sutherland, *We Harbor No Evil Design*, 57; Victor L. Settle, "Halifax Shipyards, 1918–1978: An Historical Perspective" (MA thesis, Saint Mary's University, 1994). An earlier marine railway on the Dartmouth side of the harbour predated the graving yard by nearly thirty years.

85 Raddall, *The Nymph and the Lamp*, 248.

86 "If cities embody the Enlightenment dream of controlling nature through science and technology, this dream is facilitated by all-too-tangible infrastructure. While this infrastructure is sometimes elevated as a symbol of collective achievement, providing powerful ideological legitimation for states and nations, it is frequently taken for granted, its success measured by ubiquity and invisibility." Ashley Dawson, *Extreme Cities: The Peril and Promise of Urban Life in the Age of Climate Change* (Verso Press, 2017), 130–31. On sediment and the Anthropocene, see Bruce Wilkinson, "Humans as Geologic Agents: A Deep-Time Perspective," *Geology* 33, 3 (2005): 161–64.

87 Regional leadership was caught between the need for federal funds and anxiety over what a greater Dominion presence might entail. As the mayor of Halifax told city council, the Royal Commission on Maritime Claims had spoken "in the public interest of Canada," not "in the interest of the Atlantic ports." Also the words "in the interest of the future growth and expansion of its (Canada's) activities," not of the "activities of the port of Halifax." See "Council Votes in Favour of Harbour Commission," *Evening Mail*, 3 December 1927, 3, 12; "South Terminals Scene of Great Activity," *Evening Mail*, 21 June 1928, 15. On the Maritime Rights movement see Ernest R. Forbes, *The Maritime Rights Movement, 1919–1927: A Study in Canadian Regionalism* (McGill-Queen's University Press, 1979). Cities including Vancouver, Toronto, and Montreal already had harbour commissions, most of which were folded into a National Harbours Board in 1936.

88 Schwinghamer and Raska, *Pier 21*, 45–46. See also Schwinghamer's "'Altogether Unsatisfactory': Revisiting the Opening of the Immigration Facility at Halifax's Pier 21," *Journal of the Royal Nova Scotia Historical Society* 15 (2012): 61–74.

89 Raddall, *The Nymph and the Lamp*, 333.

90 Raddall, *Halifax: Warden of the North*, 234, 268.

91 Gordon Stevenson, *A Redevelopment Study of Halifax, Nova Scotia* (1957), x, 60–61, HMA. On the context and impact of the report, see Jill Grant and Marcus Paterson, "Scientific Cloak/Romantic Heart: Gordon Stephenson and the Redevelopment Study of Halifax, 1957," *Town Planning Review* 83, 3 (2012): 319–36; Ted Rutland, *Displacing Blackness: Power, Planning, and Race in Twentieth-Century Halifax* (University of Toronto Press, 2018); and Tina Loo, *Moved by the State: Forced Relocation and Making a Good Life in Postwar Canada* (UBC Press, 2019). Urban renewal in Halifax generally focused on a district marked as the "Old Northern Suburb," reaching from the Citadel to North Street, and from Barrington Street to Agricola Street; designated the "Uniacke Urban Renewal Area" in 1962, and the "Neighbourhood Improvement Area" in 1976. H.J. Porter and Associates, *NIP Conceptual Plan Report: Halifax Neighbourhood Improvement Area* (1976).

92 *Halifax Waterfront Project, Halifax, Nova Scotia: Restoration, Proposal and Specifications* (Department of Indian & Northern Affairs, 1972); Susan Buggey, *Halifax Waterfront Buildings* (1974); *Halifax Waterfront Development Plan* (1976).

93 This had been the "Central Victual Depot" for the Department of Defence, which had occupied this part of the waterfront since the late eighteenth century. Confidential report to Mayor & City Council, 19 July 1967 from G.F. West, Acting City Manager, Department of Engineering and Works Files: Files 102-39B.916 Central Victualling Depot, 1967, HMA.

94 See correspondence in folders 102-39B.917B Central Victualling Depot, 1968, and 102-39B.918 Central Victualling Depot, 1969, HMA.

"P.S. #2 Perhaps if you were to look under the old wooden Simons building you will find the above missing fill material. P.S. #3 Isn't this act contrary to the ground rules for the restoration of historical buildings ..." D.J. Fox, Chief of Engineering Division to A.W. Churchill, Supervisor, Real Estate, 2 January 1974, Folder Historic Buildings #3, 1973–1976, 102.39B.1087; also *Halifax Waterfront Historic Buildings Engineering Study and Report,* Paul Wendt Ltd (Dartmouth), 1969; 102-39B.1086 Historic Buildings #2, 1969–73, HMA.

95 Chebucto Square, Waterfront Development Corporation Limited, "Chebucto Where????" 23 January 1978 report, Halifax Landmarks Commission Files 102.108.20, HMA. This had been granted by the Crown as a "public ground and slip granted in trust" in 1794. See [Intergovernmental Waterfront Committee,] *Halifax/Dartmouth Waterfront Plan,* February 1976 (n.p.), Folder 102-39B.1372, HMA.

96 Stevenson, *A Redevelopment Study of Halifax, Nova Scotia* (1957), x.

97 My thanks to Neil Balan for this reference.

98 Robert J. McCalla, "Separation and Specialization of Land Uses in Cityport Waterfronts: The Cases of Saint John and Halifax," *Canadian Geographer* 27, 1 (1983): 48–61; J. Philip Dumaresq, *City of Halifax 10 Year Development Plan* (24 January 1950), 6, HMA.

99 Peter Clancy, *Offshore Petroleum Politics: Regulation and Risk in the Scotian Basin* (UBC Press, 2011), 84, 144. The great irony is that after fifty years of exploration, the "Scotian Basin remains an 'almost resource,'" regardless of the ebbs and flows of capital investment or government interest. Yet hopes for fossil fuel futures, however ill-informed by history, spring eternal. See also Claire Campbell, "'Rising with the Tide of History': The Age of Sail as Industrial Alibi," *Papers in Canadian History & Environment* 2 (2019): 1–37.

100 See, for example, the Canada–Nova Scotia Offshore Petroleum Board map of licences at https://www.cnsopb.ns.ca/what-we-do/lands-management/maps-and-coordinates. Philip Steinberg, *The Social Construction of the Ocean* (Cambridge University Press, 2001), 112. Steinberg argues that the ocean proper had more capital value during the age of sail (in trade), while the shoreline became more valuable for building in industrialization.

101 Irving was a relative latecomer to the downtown harbour (1936), but remained the most prominent reminder of the fuel history with tanks until 1982 (email communication with Carolyn Gilbert, Waterfront Development Corporation, 25 May 2018). The Heritage Trust petitioned to protect the Irving Arch in the early 1980s – the last waterfront arch extant – but it was removed. Sackville Landing is now home to a provincial tourism visitors bureau, a playground, the Maritime Museum of the Atlantic, the historic HMCS *Sackville*, various memorials, and *The Wave*, a sculpture by Donna Hiebert. The façade of the Imperial Oil building was kept in the "Waterside Centre" (2012) by the same development group behind Queen's Marque.

102 Matthew McClearn, "Halifax's Battle of the Rising Sea: Will the City Be Ready for Future Floods and Storms?," *Globe and Mail*, 5 March 2018, https://www.theglobeandmail.com/news/national/rising-sea-levels-halifax/article38160929/.

In 1974 Halifax City Council adopted a bylaw that identified ten crucial "view planes" – sight lines from Citadel Hill to different points around the harbour that had been necessary to communicate with ships – as a way to protect its intangible heritage: a distinct sense of place and a reminder of the city's founding raison d'être. The views are, technically, still protected by the city's bylaws, although exemptions are allowed. The Nova Centre is not the only project (both before and since 1974) to protrude into the view planes, but it is the bulkiest. See Simpson McLeod, "Protecting the Visual Relationship Between Halifax Harbour and the Urban Area," in *Preserving the Environment of Halifax Harbour*, ed. Andre Ducharme and Gary Turner (Fisheries and Oceans Canada, 2001); and Jill L. Grant and Gladys Wai Kwan Leung, "Disputing the Character of the City: Heritage, Regeneration and the Urban Design Turn," *London Journal of Canadian Studies* 31, 1 (2016): 113–30.

As in Charlottetown, a provincial crown corporation, the Waterfront Development Corporation, was created in 1976 tasked with rehabilitating waterfront as real estate. In the kind of rhetorical framing in which historians delight, the Waterfront Development Corporation was renamed Develop Nova Scotia in 2018 and, even more on the nose, renamed again Build Nova Scotia in 2022.

Postscript: Inheritance and Imagination

1 Clarence Spencer, "Cumulative Infilling Activities," paper presented at Preserving the Environment of Halifax Harbour Workshop #2, Halifax, January 2001 (Fisheries and Oceans Canada and Halifax Regional Municipality, 2001), 73. This was the second of two meetings in the context of the "Halifax Harbour Solutions" project (1997–2010) designed to address sewage and pollution in the harbour.

2 Han Meyer, "Toward a Cultural Heritage of Adaptation: A Plea to Embrace the Heritage of a Culture of Risk, Vulnerability and Adaptation," in *Adaptive Strategies for Water Heritage: Past, Present and Future*, ed. Carola Hein (Springer, 2019). Historians have described this with different metaphors, from geological sediments to layer cakes to "old houses that have been painted many times. Each new layer of paint covers those below, making the past invisible. Yet the hidden layers persist." Matthew M. Booker, *Down by the Bay: San Francisco's History Between the Tides* (University of California Press, 2013), 12.

3 Forgive the quoting of Anne Shirley here.

4 Phyllis R. Blakeley, "Charles Morris," *Dictionary of Canadian Biography Online*, vol. 9.

5 Elizabeth Lunstrum and Pablo S. Bose, "Environmental Displacement in the Anthropocene," *Annals of the American Association of Geographers* 112, 3 (2022): 644–53. From New England

"Planters" in the 1760s, then settlers from Yorkshire to the Tantramar region, and tens of thousands of Loyalists after the Revolutionary War. Elizabeth Mancke, "Spaces of Power in the Early Modern Northeast," *New England and the Maritime Provinces: Connections and Comparisons,* ed. Stephen Hornsby and John Reid (McGill-Queen's University Press, 2005), 44–45.

6 British Museum Additional Manuscripts 19071, folio 18 (Rev. Andrew Brown collection), Paul Mascarene, "Journal of a Voyage from Annapolis Royal to Canso," 1721, courtesy NSARM; also *A Geographical history of Nova Scotia* (London, 1749), 29. See Robert Shears, "Examination of a Contested Landscape: Archaeological Prospection on the Eastern Shore of Nova Scotia" (MA Thesis, St. Mary's University, 2013).

7 Lunenburg Township land grant Crown Land Old Book 13, 208–17, 16 September 1784, NSARM. My thanks to the reference team at the Nova Scotia Archives for supplying this document. It's worth noting the grant reserved to the Crown "Mines of Gold & Silver, Lead, Copper & Coals" (page 213).

8 John Brittain, "Description of the Settlements in the Provinces of Nova Scotia & New Brunswick, 1784–1785" (William L. Clements Library M979, University of Michigan, 1955), 71–72; also *The present state of Nova Scotia* (Edinburgh: William Creech; sold by T. Cadell, and G. Robinson, London, 1787), 137–38. On the provisioning of Lunenburg, I thank Patrick Callaway, "Imperial Investments and Provincial Inducements: The Problem of Nova Scotia in the Post-Seven Years War Empire," paper presented at the Atlantic Canada Studies Conference, University of Maine, May 9–11, 2024.

9 *The American Geography, or, A View of the Present Situation of the United States of America* (Elizabeth Town, 1789), 475.

10 *Lunenburg, Nova Scotia: the Gloucester of Canada* (published for J.M. Anderson and J.J. Kinley, 1906?) V/F V.365 #17, NSARM.

11 H.W. Hewitt, "Lunenburg: The Upbuilding of the Town," *The Suburban* (1907), F 107 L97 H49, NSARM.

12 According to the census, in 1901 the town's population was 2,916; the 1961 census shows a population of 3,012.

13 "An Invitation from the Town of Lunenburg," 197? V/F V.95 #4 [Industrial Commission], NSARM.

14 Canadian–British Engineering Consultants and Norman Pearson, *Town of Lunenburg, Nova Scotia: Urban Renewal Study* (Rolph, Clark, Stone, Eastern Limited, 1966), 1, 3–4, 40, 43, 49, 51.

15 Cheryl Sullivan, "The Paradox of Bluenose II: Antimodernism, Capitalism and the Legacy of the Schooner Bluenose in Nova Scotia," *Nova Scotia Historical Review* 16, 2 (1996): 1–22. On this period of history-inflected development, see Meaghan Beaton, *The Centennial Cure: Commemoration, Identity, and Cultural Capital in Nova Scotia During Canada's 1967 Centennial Celebrations* (University of Toronto Press, 2017).

16 R. Alexander Hunter, "Cod, Colonialism, and the Anthropocene," *Environment & History* 28, 3 (2022): 369–74.

17 World Heritage Committee, Decision, *Report of the 19th Session of the Committee* (1995), and International Council on Monuments and Sites, *Advisory Body Evaluation* (1995). UNESCO World Heritage designation can include recognition of "traditional human settlement, land-use, or sea-use which is representative of a culture (or cultures), or human interaction with the environment especially when it has become vulnerable under the impact of irreversible change."

18 Claire Campbell, "Global Expectations, Local Pressures: Some Dilemmas of a World Heritage Site," *Journal of the Royal Nova Scotia Historical Society* 11, 1 (2008): 69–88.

19 Haley Ryan, "Council Paves Way for Sale of Lunenburg Land That Could See 256 New Homes," *CBC News*, 11 April 2024, https://www.cbc.ca/news/canada/nova-scotia/lunenburg-declares-public-land-surplus-plans-for-250-new-homes-1.7170033; Haley Ryan, "Lunenburg Town Council Halts Development of Municipal Lands," *CBC News*, 17 December 2025, https://www.cbc.ca/news/canada/nova-scotia/lunenburg-town-council-halts-development-municipal-lands-9.7018515. This project would still keep the town's population below 1871 figures (1871: 3,231; 2021: 2,396) but would represent a significant addition (i.e., approximately +30 percent).

20 "Description of the Island, October 5, 1765," reprinted in the *Journal of the House of Assembly of Prince Edward Island, Fourth Session of the Fifteenth General Assembly, Appendix D* (Cooper & Bremner, Charlottetown, 1841), 11–13. For a detailed local history, see L. Currie and B. Mair, *Georgetown, Prince Edward Island, 1730–1900* (Georgetown Lions Club, 1982).

21 John Stewart, *An Account of Prince Edward Island in the gulf of St. Lawrence* (London: Winchester & Son, 1806), 17–20.

22 The *Census of the population of Prince Edward Island, taken under the authority of the act Geo. IV. c. 4, Anno Domini, 1827* counts 156 for Georgetown and Royalty, 1,649 for Charlottetown (plus 424 for Charlottetown Royalty), and 314 for Princetown Royalty.

23 Walter Johnstone, *A Series of Letters, Descriptive of Prince Edward Island in the Gulph of St. Laurence* (Dumfries: J. Swan, 1822), 32.

24 *Hutchinson's Prince Edward Island directory, for 1864: containing an alphabetical directory of the citizens, also a classified business directory, and an appendix containing much useful information.*

25 Brian Payne, "'The Best Fishing Station': The Fish Trade of Prince Edward Island and Resource Transfer in the Gulf of St Lawrence, 1854–73," in *The Greater Gulf: Essays on the Environmental History of the Gulf of St. Lawrence*, ed. Claire Campbell, Edward MacDonald, and Brian Payne (McGill-Queen's University Press, 2019).

26 Nicolas J. De Jong and Marven E. Moore, *Shipbuilding on Prince Edward Island: Enterprise in a Maritime Setting, 1787–1920* (Canadian Museum of Civilization, 1994), 162–63, 291–304.

27 Birch Bagster, *Progress and Prospects of Prince Edward Island, written during the leisure of a visit in 1861* (Charlottetown: Queen's Printer, 1861), 44, 58–59.

28 Neil Matheson, "Georgetown and Royalty," *Abstract of the Census of the Population, and Other Statistical Returns, of Prince Edward Island, taken in the year 1871* (Charlottetown, 1871), n.p.

29 G. Edward MacDonald, "Andrew Archibald Macdonald," *Dictionary of Canadian Biography*, vol. 14 (University of Toronto/Université Laval, 2003–); and "The Macdonalds of Panmure Island and the Challenge of Change on Prince Edward Island, 1805–1919," *American Review of Canadian Studies* (2025).

30 *Beautiful Prince Edward Island: the summer land of North America* (n.p., 1893), 6.

31 *The garden of the gulf, Prince Edward Island: and its handsome and delightful capital Charlottetown: being pictures and description of the charms of city and seashore as summer resorts* (W.G. McFarlane/Geo. Carter, 1900), 21; see Alan MacEachern and George Edward MacDonald, *The Summer Trade: A History of Tourism on Prince Edward Island* (McGill-Queen's University Press, 2022).

32 G. Edward MacDonald, *If You're Stronghearted: Prince Edward Island in the Twentieth Century* (Prince Edward Island Museum and Heritage Foundation, 2000), 268.

33 *Report of the Commission of Inquiry into Matters Pertaining to Bathurst Marine Limited and Gulf Garden Foods Limited* (July 1969), 12. The report follows the labyrinthine business history of the Georgetown projects of the 1960s (with some wry commentary). By this point the province had sunk about $9.5 million into the three main assets – fish plant, shipyard, and a few trawlers – which, the commission concluded, were "worth much less than what was spent to acquire them" (71).

34 Amid the backlash, one political scientist argued that "this policy of forced growth would soon entangle Islanders, as it had Canadians elsewhere, in the dismal politics of luring in outside businesses with generous concessions and of providing outlays of public funds to keep them afloat. It was also introducing an alternate vision of development on the Island, based more and more on the standard industrial model." David Milne, "Politics in the Garden," in *The Garden Transformed: Prince Edward Island, 1945–1980,* ed. Verner Smitheram, David Milne, and Satadal Dasgupta (Ragweed Press, 1982), 47; Macdonald, *If You're Stronghearted,* 273–75. One might respond that Canada has a long, proud, and thriving tradition of subsidizing corporate projects with the public purse, but it was clear that *Islanders* did not feel invested in this project, despite its ostensible connection with Georgetown's past. The project initially proposed a salmon farm as well, somewhat ahead of its time, given the development of aquaculture on both coasts in recent years. Errol Sharpe provides a scathing critique of the "fiasco" in *A People's History of Prince Edward Island* (Steel Rail Publishing, 1976), 199–201.

35 Robert C. Tuck, "Georgetown: The Town That Time Forgot," *The Island Magazine*, 4 (Spring/Summer 1978): 29.

36 The 2021 census shows a population of 544; in 1901, it was 1,123. The *Abstract of the census of the population and other statistical returns of Prince Edward Island taken in the year 1848* showed a population of 556.

37 "Breaking Ground at Georgetown Clean Tech Park," 15 February 2023, https://www.princeedwardisland.ca/en/news/breaking-ground-at-georgetown-clean-tech-park. Three Rivers was created in 2018 by amalgamating Georgetown with the larger town of Montague plus five rural municipalities and eighteen rural areas.

38 Stephen Patterson, "Eighteenth-Century Treaties: The Mi'kmaq, Maliseet, and Passamaquoddy Experience," *Native Studies Review* 18, 1 (2009): 40; John Reid, "Empire, the Maritime Colonies, and the Supplanting of Mi'kma'ki/Wulstukwik, 1780–1820," *Acadiensis* 38, 2 (2009): 85.

39 On the boundary question, see David Demeritt, "Representing the 'True' St. Croix: Knowledge and Power in the Partition of the Northeast," *William & Mary Quarterly* 54, 3 (1997); Francis M. Carroll, *A Good and Wise Measure: The Search for the Canadian-American Boundary, 1783–1842* (University of Toronto Press, 2001); and Michah Pawling, ed., *The 1820 Journal and Plans of Survey of Joseph Treat* (University of Massachusetts Press, 2007). The Maine/New Brunswick border (and other contentious and unclear sections of the border established in 1783) would not be resolved until the Webster-Ashburton Treaty of 1842. Stephen J. Hornsby, "Negotiating the International Boundary," Plate 21, in *Historical Atlas of Maine.*

40 Nineteen thousand acres, according to Richard Diubaldo, "The Early History of St. Andrews, New Brunswick, 1784–1814," in *St. Andrews, N.B.: Miscellaneous Reports,* Manuscript Report Number 134 (Parks Canada, 1965–66), 92.

41 "Representation of the Inhabitants of Saint Andrews," 26 May 1784, Edward Winslow Family Papers, University of New Brunswick.

42 Joshua Smith, *Borderland Smuggling: Patriots, Loyalists, and Illicit Trade in the Northeast, 1783–1820* (University Press of Florida, 2006). For an accessible account of the economic life of Saint Andrews as a border town and shipping port, affected by war and different eras of protectionism, see Ronald Rees, *St. Andrews and the Islands* (Nimbus Publishing, 1995).

43 "Saint John to be the Eastern terminus, St. Andrews May Become a Coal Depot," *St. Andrews Beacon,* 21 August 1890, reproduced in David Sullivan, ed., *St. Andrews: An Historical Scrapbook* (Pendlebury Press, 2008), 98–99. In a labyrinthine evolution characteristic of nineteenth-century railway construction in British North America, the railway into Saint Andrews had first been chartered in 1836 as the St. Andrews and Quebec Railway; folded into the New Brunswick and Canada Railway and, in 1882, the New Brunswick Railway Company; and in 1890 acquired by the Canadian Pacific Railway.

44 See, for example, for Ontario, Claire Campbell, *Shaped by the West Wind: Nature and History in Georgian Bay* (UBC Press, 2004); and Andrew Watson, *Making Muskoka: Tourism, Rural Identity, and Sustainability, 1870–1920* (UBC Press, 2023).

45 *St. Andrew's, New Brunswick on Passamaquoddy Bay reached by "International" steamers from Boston or Portland* (Boston: Rand Avery, 1889), 5; *Toward the sunrise: A guide to the seacoast resorts of eastern Maine, New Brunswick, Nova Scotia, Prince Edward Island and Cape Breton* (Boston: International Steamship Company, 1895), 31; "Great Movement in Real Estate," *The Bay Pilot,* 1 March 1888, 2. To be fair, the Land Company was presided over by Samuel Tilley.

46 *The Algonquin, St. Andrews, N.B.: The first season's verdict, regarding St. Andrews and the Algonquin, confirms all that had been said and written* (n.p., 1890). The town newspaper worried that its streetscape might be a bit *too* rustic for the Company's urbane aspirations: "The final ornamentation of the Park and Park drives and walks has been and must be delayed until the running at large of cows is put a stop to." "Cows at Large," *St. Andrews Bay Pilot,* 23 August 1888, 2.

47 Elizabeth L. Jewett, "What Was Driving Golf? Mobility, Nature, and the Making of Canadian Leisure Landscapes, 1870–1930," in *Moving Natures: Mobility and the Environment in Canadian History,* ed. Ben Bradley, Jay Young, and Colin M. Coates (University of Calgary Press, 2016).

48 *The Algonquin, St. Andrews, N.B.* (n.p., 1892).

49 Canadian Pacific Railway Company, *St. Andrews by-the-Sea, New Brunswick: Reached by the Canadian Pacific Railway* (1900), 1.

50 *St. Andrew's, New Brunswick on Passamaquoddy Bay* (1889), 11; Canadian Pacific Railway Company, *St. Andrews by-the-Sea,* 2; *The Algonquin, St. Andrews, N.B.: the first season's verdict* (1890); Andrew Sackett, "Inhaling the Salubrious Air: Health and Development in St. Andrews, N.B., 1880–1910," *Acadiensis* 25, 1 (1995): 54–81.

51 Eric Mills, "Science in Canada: The Context of the Biological Board of Canada's St Andrews Biological Station," and Robert L. Stephenson, "St Andrews Biological Station: A Case Study in the History of Public Science in Canada," in *A Century of Maritime Science: The St. Andrews Biological Station,* ed. Jennifer Hubbard, Robert L. Stephenson, and David J. Wildish (University of Toronto Press, 2016).

52 George C. Ingram, "The West Blockhouse St. Andrews, N.B.; A Structural Report," in *St. Andrews, N.B.: Miscellaneous Reports,* Manuscript Report no. 134 (Parks Canada, 1965–66). Historic districts designated in this period included Water Street, St. John's (1987); Great George Street, Charlottetown (1990); Lunenburg (1991); Annapolis Royal, Nova Scotia (1994); Hydrostone District, Halifax (1994); and Grand Pré (1995).

53 Census counts for 1881 show a population of 2,128; for 2021, a population of 2,048. A majority of dwellings are occupied by what the census calls "a usual resident" (i.e., year-round) – and the town itself has seen an increase in this figure from 2016 (79 percent) to 2021 (84 percent). My thanks to Paul Knopper (senior administrator, Town of St. Andrews) and Xander Gopen (Planner/Housing Specialist) for this information. The town saw roughly ten thousand visitors in 2023, according to the Town of Saint Andrews Regular Council Meeting Minutes, 5 September 2023.

54 "Storm Surges Force 'Revamp' of Downtown Saint Andrews," *CBC News,* 3 October 2017, https://www.cbc.ca/news/canada/new-brunswick/st-andrews-storm-climate-change-1.4318845. Parks Canada reports in the 1960s (see Ingram, note 52) had already expressed concerns about erosion with high tides washing against the blockhouse and battery.

55 *Saint Andrews Community Based Climate Adaptation Plan* (2019); Town of Saint Andrews Regular Council Meeting Minutes, 15 January 2024.

56 Renata Tyszczuk, *Provisional Cities: Cautionary Tales for the Anthropocene* (Routledge, 2019), 7.

57 St. Andrews Blockhouse National Historic Site of Canada Heritage Character Statement, FHBRO Number 98-120, n.d.; Statement of Significance, St. Andrews Historic District National Historic Site of Canada, Parks Canada.

58 "Nature in Urban Saint John," *Journal of the New Brunswick Museum* (1978): 80.

59 Bagster, *Progress and Prospects*, 59.

60 My thanks to Edward MacDonald for his tour of Georgetown and his thoughts on this question, 25 June 2023.

61 H.W. Hewitt, "Lunenburg: The Upbuilding of the Town (from the Suburban)," 1907, F 107 L97 H49, NSARM.

62 Graeme Wynn, "Reflections on the Environmental History of Atlantic Canada," in *Land and Sea: Environmental History in Atlantic Canada,* ed. Claire Campbell and Robert Summerby-Murray (Acadiensis Press, 2013), 254. David Alexander likewise suggested that Newfoundland might do better to act more *as* an island, as bounded and finite, with "prosperity" defined less by the quantitative measures from other places. "Newfoundland's Traditional Economy and Development to 1934," in *Newfoundland in the Nineteenth and Twentieth Centuries: Essays in Interpretation,* ed. James Hiller and Peter Neary (University of Toronto Press, 1980), 36–37.

63 Isaac Land, "The Urban Amphibious," in *The New Coastal History: Cultural and Environmental Perspectives from Scotland and Beyond,* ed. David Worthington (Palgrave Macmillan, 2017), 40.

As examples of creative framings, see Michael H. Carriere and David Schalliol, *The City Creative: The Rise of Urban Placemaking in Contemporary America* (University of Chicago Press, 2021); Maurizio Carta and Daniele Ronsivalle, eds., *The Fluid City Paradigm: Waterfront Regeneration as an Urban Renewal Strategy* (Springer, 2016); and Carola Hein, "Planning for Porosity: Exploring Port City Development Through the Lens of Boundaries and Flows," *Urban Planning* 6, 3 (2021).

64 Research into waterfront regeneration does tend to be biographical (per an individual city) or regional; for example, Harry Smith and Maria Soledad Garcia Ferrari, "Lessons from Shared Experiences in Sustainable Waterfront Regeneration Around the North Sea," in *Waterfront Regeneration: Experiences in City-Building*, ed. Harry Smith and Maria Soledad Garcia Ferrari (Routledge, 2012).

Bibliography

Archives

Library and Archives Canada
Halifax Regional Municipality Archives
Dalhousie University Archives and Special Collections
Archives of Nova Scotia, Halifax
Provincial Archives of New Brunswick
New Brunswick Museum
Public Archives and Records Office of Prince Edward Island, Charlottetown
Robertson Library, University of Prince Edward Island
St. John's City Archives, Newfoundland
Queen Elizabeth Library, Memorial University, Archives and Special Collections/Centre for Newfoundland Studies
William L. Clements Library, University of Michigan

Newspapers and Media

Globe and Mail
Montreal Gazette
Nova Scotia: *Acadian Recorder, Daily Citizen, Daily Echo, Halifax Evening Mail, Halifax Herald, Morning Chronicle, Nova Scotian*
Prince Edward Island: *The Examiner, The Guardian, Haszard's Gazette*
New Brunswick: *The Royal Gazette and the New-Brunswick Advertiser, St. Andrews Bay Pilot*
Newfoundland: *Daily Tribune, Evening Telegram, Morning Courier, The Patriot*
Canadian Broadcasting Corporation

Legislative and Executive Records

House of Commons *Debates*
Journal of the House of Assembly of Newfoundland
Journal of the Legislative Council, Newfoundland
Journal of the House of Assembly of Prince Edward Island
Journal of the House of Assembly of the Province of New Brunswick
Statutes of Newfoundland
Censuses, various.
Copies Or Extracts of Correspondence Relative to the Affairs of British North America, Colonial Office, 1839. Memorial University.
Copies or extracts of the correspondence between the Governor of Newfoundland and the Secretary of State for the Colonies, in reference to the appropriation of the subscriptions raised for the relief of the sufferers at St. John's by the fire in 1846. Memorial University.
Despatches from the Governor of Newfoundland to the Secretary of State [for the Colonies], vol. 1, 1800–1813 and *Despatches received by the Governor of Newfoundland from the Secretary of State for the Colonies, vol. 2, 1814–1819*. Memorial University.
Laws and ordinances, ordained and established by the mayor, aldermen and commonalty of the city of New-York, in Common Council convened. Samuel and John Loundon, 1786.
Letter to the Earl of Carnarvon by Mr. Joseph Howe, Mr. William Annand, and Mr. Hugh McDonald Stating their Objections to the Proposed Scheme of Union of the British North American provinces, Presented to both Houses of Parliament. London: Eyre and Spottiswoode, 1867.
The charters of the city of London, which have been granted by the kings and queens of England since the conquest. London: J. Hodges, 1745.
The historical charters and constitutional documents of the city of London. London: Whiting, 1884.

Printed Primary Sources

A Sketch of the state of affairs in Newfoundland By A Late Resident in that Colony. London: Sauders & Otley, 1841.
Akins, Thomas B. *History of Halifax City.* Halifax, n.p., 1895.
Allen, William Ferneley. *The Corporation of London: Its Rights and Privileges.* London: William H. Allen, 1858.
Anspach, Lewis. *A history of the island of Newfoundland, containing a description of the island, the banks, the fisheries and trade of Newfoundland, and the coast of Labrador.* London: T. and J. Allman and J.M. Richardson, 1819.
Bagster, Birch. *Progress and Prospects of Prince Edward Island, written during the leisure of a visit in 1861.* Charlottetown: Queen's Printer, 1861.
Bell, Frederick McKelvey. *A Romance of the Halifax Disaster*. The Royal, 1918.
Bellin, Jacques-Nicholas. "Remarks by M. Bellin in Relation to His Maps Drawn for P. Charlevoix's History of New France," *The Gentleman's Magazine*, February 1746.
Brogan, Bailey, Jamylynn McDonald, Blaine Lyons, Peter Johnston, and Graeme Stewart-Robertson. *Climate Change Adaptation Plan for Saint John.* Atlantic Coastal Action Program, 2020.
Burchell, Herbert C. "The colonial government dry dock, St. John's Newfoundland." *Transactions of the Canadian Society of Civil Engineers*, 1887.

Campbell, Duncan. *History of Prince Edward Island.* Charlottetown: Bremner Bros., 1875.
Canada's First City – Saint John, N.B.: The Charter of 1785 and Common Council Proceedings. Lingely Printing, 1962.
Canadian–British Engineering Consultants and Norman Pearson. *Town of Lunenburg, Nova Scotia: Urban Renewal Study.* Rolph, Clark, Stone, Eastern Limited, 1966.
Cartier, Jacques. *Brief recit & fuccincte narration, de la nauigation faicte es yiles de Canada, Hochelage & Saguenay & autres.* A. Le Clerc/P. Roffet, 1545. Available at Bibliothèque Nationale de France, https://gallica.bnf.fr/ark:/12148/bpt6k521342.
Charlottetown Area Development Corporation. *Environmental Impact Statement for Concert Venue Site and Parking Facility, Imperial Oil Co. and Adjacent Lands.* 2010.
Collections of the Massachusetts Historical Society. "Miscellaneous remarks and observations on Nova Scotia, New Brunswick and Cape Breton, Supposed to be written by the Surveyor General of Nova Scotia." March 1794.
Collections of the New Brunswick Historical Society. Assorted papers.
Colton's Illustrated Cabinet Atlas and Descriptive Geography. New York, 1859.
Cooney, Robert. *The Autobiography of a Wesleyan Methodist Missionary.* Montreal, 1856.
Cowie, Frederick William. "Report on the River St. Lawrence Ship Channel from Montreal to Quebec and the Traverse." Government Printing, 1906.
Cowie, Frederick William. *Report to the Honourable Frank Cochrane, Minister of Railways and Canals, on Halifax Harbour and the development of a project of modern ocean terminals.* Montreal, 1913.
Cowie, Frederick William. "Transportation Routes in Canada." *The Journal of the Engineering Institute of Canada* 6, 7 (July 1923): 313–22.
Ellis, Henry, ed. *The new chronicles of England and France, in two parts: by Robert Fabyan. Named by himself The concordance of histories, ca 1516.* London: F.C. & J. Rivington, 1811.
Eveleth, Stanford. *Miss Dexie: A Romance of the Provinces.* Toronto: W. Briggs, 1895.
Fenety, George E. *Political notes and observations, or, A glance at the leading measures that have been introduced and discussed in the House of Assembly of New Brunswick.* Fredericton: S.R. Miller, 1867.
Field, Cyrus. "Statement of some of the advantages attendant upon making St. John's, Newfoundland a port of call for transatlantic steamers London." London: M. Lounds, 1856.
Fisher, Peter. *Notitia of New-Brunswick, for 1836, and Extending Into 1837.* London: H. Chubb, 1838.
Fisher, Peter. *Sketches of New-Brunswick: containing an account of the first settlement of the province: with a brief description of the country, climate, productions, inhabitants, government, rivers, towns, settlements, public institutions, trade, revenue, population &c.* Saint John: Chubb & Sears, 1825.
Gesner, Abraham. *First Report on the Geological Survey of New Brunswick.* Saint John: H. Chubb, 1839.
Gesner, Abraham. *New Brunswick; with notes for emigrants.* London: Simmonds & Ward, 1847.
Giles, Jacob. *A new law-dictionary.* London: In the Savoy, 1732.
Haliburton, Thomas. *A general description of Nova Scotia.* Halifax: Clement Belcher, 1825.
Halifax Waterfront Development Plan. 1976.
Halifax Waterfront Project, Halifax, Nova Scotia: Restoration, Proposal and Specifications. Department of Indian & Northern Affairs, 1972.
Harvey, Moses. *Text-book of Newfoundland history, for the use of schools and academies.* Boston: Doyle and Whittle, 1885.

Hatton, Joseph, and Moses Harvey. *Newfoundland: The oldest British colony, its history, its present condition, and its prospects in the future.* London: Chapman and Hall, 1883.

Information to emigrants: an account of the island of Prince Edward, with practical advice to those intending to emigrate ... By Late resident of that colony. London: James Asperne, 1819.

Johnston, J.F.W. *Notes on North America.* London: Blackwood and Sons, 1851.

Johnstone, Walter. *A Series of Letters, Descriptive of Prince Edward Island in the Gulph of St. Laurence.* Dumfries: J. Swan, 1822.

Keating, E.H., City Engineer. "Preliminary Report on the Halifax Dry Dock and Report on American Stone and Wooden Docks." Halifax: Nova Scotia Printing, 1883.

Keefer, Thomas H. *Report upon the harbor of Saint John, New Brunswick.* Saint John: Barnes and Company, 1861.

Lawson, John. *Letters on Prince Edward Island.* Charlottetown: G.T. Haszard, 1851.

Lockwood, Anthony. *Report on the projected canal across the istmus [sic] that divides Nova-Scotia and New-Brunswick: explored and levelled in the autumn of 1819.* n.p., n.d.

Marsh, George Perkins. *Man and Nature: Or, Physical Geography as Modified by Human Action.* New York: Charles Scribner, 1864.

Minutes of the Proceedings of the Institution of Civil Engineers. Assorted.

Morris, Isaac C. *The dawn of the twentieth century, as pertaining to Newfoundland.* St. John's: G.S. Milligan, 1899.

Morse, Jedidiah. *The American Geography, or, A View of the Present Situation of the United States of America.* Elizabeth Town: Shepard Kollock, 1789.

NIP Conceptual Plan Report: Halifax Neighbourhood Improvement Area. H.J. Porter & Associates, 1976.

Noble, Louis Legrand. *After icebergs with a painter: A summer voyage to Labrador and around Newfoundland.* New York: D. Appleton, 1861.

Noorthouck, John. *A New History of London Including Westminster and Southwark.* London: R. Baldwin, 1773.

Norton, George. *Commentaries on the history, constitution, and chartered franchises of the city of London.* London: Longmans Green, 1869.

Parker, C.W. *Who's Who and Why: A Biographical Dictionary of Men and Women of Canada and Newfoundland.* International Press, 1916.

Pedley, Charles. *The history of Newfoundland from the earliest times to the year 1860.* Longman, Roberts & Green, 1863.

Perley, Moses H. *On the Early History of New Brunswick: A portion of a lecture delivered before the Mechanics' Institute, St. John, in 1841 ... with notes by W.F. Ganong.* Saint John: The Educational Review, 1891.

The present state of Nova Scotia. Edinburgh: William Creech; sold by T. Cadell, and G. Robinson, London, 1787.

"Power from Tidal Currents in Bay of Fundy." *Journal of the Royal Society of Arts* 64, 3333 (1916): 788.

Prowse, D.W. *A history of Newfoundland from the English, colonial, and foreign records.* London: Macmillan, 1895.

Rae, William Fraser. *Newfoundland to Manitoba, through Canada's maritime, mining, and prairie provinces.* New York: G.P. Putnam's Sons, 1881.

Report of the Commission of Inquiry into Matters Pertaining to Bathurst Marine Limited and Gulf Garden Foods Limited. Charlottetown, July 1969.

Report of the commissioners appointed to examine into the present condition of the harbour of Saint John, and to report the same to His Excellency the Lieutenant-Governor of the province: with various documents relating thereto. Saint John, 1839.
Saint Andrews Community Based Climate Adaptation Plan. 2019.
Sherlock, R.L. *Man as a Geological Agent.* H.F. & G. Witherby, 1922.
Sherlock, R.L. "The Influence of Man as an Agent in Geographical Change." *The Geographical Journal* 61, 4 (1923): 258–68.
Stevenson, Gordon. *A Redevelopment Study of Halifax, Nova Scotia.* Corporation of the City of Halifax, 1957.
Stewart, John. *An Account of Prince Edward Island in the gulf of St. Lawrence.* London: Winchester & Son, 1806.
Wallace. F.W. *Wooden Ships and Iron Men: The Story of the Square-Rigged Merchant Marine of British North America, the Ships, Their Builders and Owners, and the Men Who Sailed Them.* Hodder and Stoughton, 1924.
Warburton, Eliot. *Hochelaga: Or, England In the New World.* London: Henry Coburn, 1846.

Directories

Frederick's Prince Edward Island directory and book of useful information for 1889–90.
Hutchinson's Prince Edward Island directory, for 1864: containing an alphabetical directory of the citizens, also a classified business directory, and an appendix containing much useful information. Charlottetown, 1864.
McAlpine Halifax City Directory, 1916.

Pamphlets, Ephemera

The Algonquin, St. Andrews, N.B. 1892.
The Algonquin, St. Andrews, N.B.: The first season's verdict, regarding St. Andrews and the Algonquin, confirms all that had been said and written. 1890.
Beautiful Prince Edward Island: the summer land of North America. n.p., 1893.
The garden of the gulf, Prince Edward Island: and its handsome and delightful capital Charlottetown: being pictures and description of the charms of city and seashore as summer resorts. W.G. McFarlane/Geo. Carter, 1900.
A Guide to the city of Halifax, Nova Scotia, 1884 & 1885: Its past and present history. Halifax: William McNab, 1884.
Halifax - Nova Scotia: The Garrison City by the Sea. Nova Scotia Tourist Association, 1900.
Halifax Nova Scotia and Its Attractions. Howard and Kutsche, 1904.
Halifax: The City by the Sea. Halifax Board of Trade, ca. 1920.
The maritime provinces: a handbook for travellers. Boston: J.R. Osgood, 1875.
Saint John as a Canadian winter port and terminus of the Canadian Pacific and Intercolonial Railways. City Corporation/Board of Trade, 1898.
St. Andrew's, New Brunswick on Passamaquoddy Bay reached by "International" steamers from Boston or Portland. Boston: Rand Avery, 1889.
St. Andrews by-the-Sea, New Brunswick: Reached by the Canadian Pacific Railway. Canadian Pacific Railway Company, 1900.
St. John and its business: A history of St. John and a statement in general terms of its various kinds of business successfully prosecuted. 1875.

Toward the sunrise: A guide to the seacoast resorts of eastern Maine, New Brunswick, Nova Scotia, Prince Edward Island and Cape Breton. Boston: International Steamship Company, 1895.

Databases

British North American Legislative Database (https://bnald.lib.unb.ca).
Canada's Register of Historic Places (www.historicplaces.ca).
CHRIS: PEI Climate Hazard and Risk Information (https://chris.peiclimate.ca).

Secondary Sources

Acheson, T.W. "The National Policy and the Industrialization of the Maritimes, 1880–1910." *Acadiensis: Journal of the History of the Atlantic Region/Revue d'histoire de la Région Atlantique* 1, 2 (1972): 3–28.

Acheson, T.W. *Saint John: The Making of a Colonial Urban Community.* University of Toronto Press, 1985.

Adamson, George C.D., Matthew J. Hannaford, and Eleonora J. Rohland, "Re-thinking the Present: The Role of a Historical Focus in Climate Change Adaptation Research." *Global Environmental Change* 48 (2018): 195–205.

Ash, Eric H. *The Draining of the Fens: Projectors, Popular Politics, and State Building in Early Modern England.* Johns Hopkins University Press, 2017.

Baker, Melvin. *Aspects of Nineteenth Century St. John's Municipal History.* H. Cuff Publications, 1982.

Baker, Melvin. "The Influence of Absentee Landlordism on the Development of Municipal Government in 19th Century St. John's." *Newfoundland Quarterly* 81, 2 (1985): 19–25.

Baldwin, Douglas, and Thomas Spira, eds. *Gaslights, Epidemics, and Vagabond Cows: Charlottetown in the Victorian Era.* Ragweed Press, 1988.

Bannister, Jerry. "A River Runs Through It: Churchill Falls and the End of Newfoundland History." *Acadiensis* 41, 1 (2012): 211–25.

Bannister, Jerry. *The Rule of the Admirals: Law, Custom, and Naval Government in Newfoundland, 1699–1832.* Osgoode Society for Canadian Legal History/University of Toronto Press, 2003.

Beaton, Meaghan. *The Centennial Cure: Commemoration, Identity, and Cultural Capital in Nova Scotia During Canada's 1967 Centennial Celebrations.* University of Toronto Press, 2017.

Beaven, Brad, Karl Bell, and Robert James, eds. *Port Towns and Urban Cultures: International Histories of the Waterfront, 1700–2000.* Palgrave Macmillan, 2016.

Beckert, Sven, Ulbe Bosma, Mindi Schneider, and Eric Vanhaute. "Commodity Frontiers and the Transformation of the Global Countryside: A Research Agenda." *Journal of Global History* 16, 3 (2021): 435–50.

Bell, David. *Early Loyalist Saint John: The Origin of New Brunswick Politics, 1783–1786.* New Ireland Press, 1983.

Bell, John, ed. *Halifax: A Literary Portrait.* Pottersfield Press, 1990.

Bittermann, Rusty, and Margaret McCallum. "'One of the Finest Grass Countries I Have Met With': Prince Edward Island's Colonial-Era Cattle Trade." *Agricultural History* 90, 2 (2016): 173–94.

Bittermann, Rusty, and Margaret McCallum. "The One That Got Away: Fishery Reserves in Prince Edward Island," *Dalhousie Law Journal* 28, 2 (2005): 385–408.

Black, Brian, and Michael J. Chiarappa, eds. *Nature's Entrepôt: Philadelphia's Urban Sphere and Its Environmental Thresholds.* University of Pittsburgh Press, 2012.

Black, Fiona. "The Politics of Energy Sources." In *The Democracy Cookbook: Recipes to Renew Governance in Newfoundland and Labrador*, edited by Alex Marland and Lisa Moore. Memorial University Press, 2017.

Blake, Raymond, and Melvin Baker. *Where Once They Stood: Newfoundland's Rocky Road Towards Confederation.* University of Regina Press, 2019.

Bloomfield, G.T. "Canadian Fire Insurance Plans and Industrial Archeology." *IA: The Journal of the Society for Industrial Archeology* 8, 1 (1982): 67–80.

Bonnell, Jennifer. *Reclaiming the Don: An Environmental History of Toronto's Don River Valley.* 2nd ed. University of Toronto Press, 2024.

Bonnell, Jennifer, and Sean Kheraj. "Urban Environmental History in Anglophone Canada: Omissions and Opportunities." *Urban History Review* 50, 1–2 (2022): 45–52.

Bonneuil, Christophe. "The Geological Turn: Narratives of the Anthropocene." In *The Anthropocene and the Global Environmental Crisis*, edited by Clive Hamilton, Christophe Bonneuil, and François Gemenne. Routledge, 2015.

Booker, Matthew M. *Down by the Bay: San Francisco's History Between the Tides.* University of California Press, 2013.

Bouchier, Nancy, and Ken Cruikshank. *The People and the Bay: A Social and Environmental History of Hamilton Harbour.* UBC Press, 2016.

Bowden, Martyn J. *Great Ports of the Mercantile Era.* Edited by Stephen J. Hornsby. University of Maine Press, 2023.

Boylan, John. "The Best Laid Plans: Fire Insurance Mapping on Prince Edward Island." *Island Magazine* 58 (2005): 23–27.

Bradley, Ben, Jay Young, and Colin M. Coates, eds. *Moving Natures: Mobility and the Environment in Canadian History.* University of Calgary Press, 2016.

Brosnan, Kathleen A., Ann Durkin Keating, and William C. Barnett, eds. *City of Lake and Prairie: Chicago's Environmental History.* University of Pittsburgh Press, 2020.

Brückner, Martin, ed. *Early American Cartographies.* Omohundro Institute of Early American History and Culture, University of North Carolina Press, 2011.

Brückner, Martin. *The Social Life of Maps in America, 1750–1860.* University of North Carolina Press, 2017.

Buckner, Philip. "The Maritimes and Confederation: A Reassessment," *Canadian Historical Review* 71, 1 (1990): 1–30.

Buggey, Susan. "Building Halifax, 1841–1871." *Acadiensis* 10, 1 (1980): 90–112.

Buggey, Susan. "Halifax Waterfront Buildings: An Historical Report." Occasional Papers in Archaeology and History no. 9. National Historic Parks and Sites Branch, Parks Canada, 1974.

Bullard, Melissa Meriam, *Brooklyn's Renaissance: Commerce, Culture, and Community in the Nineteenth-Century Atlantic World.* Palgrave Macmillan, 2017.

Bumsted, J.M. *Land, Settlement, and Politics on Eighteenth-Century Prince Edward Island.* McGill-Queen's University Press, 1987.

Burtynsky, Edward. *The Industrial Sublime.* Weber State University, 2011.

Butler, David R., ed. "The Anthropocene." Special issue, *Annals of the American Association of Geographers* 111, 3 (2021).

Butler, Paul. *St. John's, City of Fire*. Flanker Press, 2007.
Buttenwieser, Ann. *Manhattan Water-Bound: Manhattan's Waterfront from the Seventeenth Century to the Present*. 2nd ed. Syracuse University Press, 1999.
Callaway, Patrick. "Imperial Investments and Provincial Inducements: The Problem of Nova Scotia in the Post-Seven Years War Empire." Paper presented at Atlantic Canada Studies Conference, University of Maine, 9–11 May 2024.
Cameron, Elspeth. *Hugh MacLennan: A Writer's Life*. University of Toronto Press, 1981.
Campbell, Claire. "Global Expectations, Local Pressures: Some Dilemmas of a World Heritage Site." *Journal of the Royal Nova Scotia Historical Society* 11, 1 (2008): 69–88.
Campbell, Claire. "Hardened Water: The Remaking of a Coastal City." *Coastal Studies & Society* 2, 1 (2022): 112–36.
Campbell, Claire. "Privileges and Entanglements: Lessons from History for Nova Scotia's Politics of Energy." *Acadiensis* 42, 2 (2013): 114–37.
Campbell, Claire. "'Rising with the Tide of History': The Age of Sail as Industrial Alibi." *Papers in Canadian History and Environment* 2, 1 (2019): 1–37.
Campbell, Claire. *Shaped by the West Wind: Nature and History in Georgian Bay*. UBC Press, 2004.
Campbell, Claire. "Whatever Happened to Pleasant Street? Rediscovering an Urban Shoreline in Halifax, Nova Scotia." *Environmental History* 25, 1 (January 2020): 134–49.
Campbell, Claire, Edward MacDonald, and Brian Payne, eds. *The Greater Gulf: Essays on the Environmental History of the Gulf of St. Lawrence*. McGill-Queen's University Press, 2019.
Campbell, Claire, and Robert Summerby-Murray, eds. *Land and Sea: Environmental History in Atlantic Canada*. Acadiensis Press, 2013.
Carp, Benjamin L. "Fire of Liberty: Firefighters, Urban Voluntary Culture, and the Revolutionary Movement," *The William and Mary Quarterly* 58, 4 (2001): 781–818.
Carriere, Michael H., and David Schalliol. *The City Creative: The Rise of Urban Placemaking in Contemporary America*. University of Chicago Press, 2021.
Carroll, Francis M. *A Good and Wise Measure: The Search for the Canadian-American Boundary, 1783–1842*. University of Toronto Press, 2001.
Carta, Maurizio, and Daniele Ronsivalle, eds. *The Fluid City Paradigm: Waterfront Regeneration as an Urban Renewal Strategy*. Springer, 2016.
Chiarappa, Michael J., and Kristin M. Szylvian. "Heeding the Landscape's Usable Past: Public History in the Service of a Working Waterfront." *Buildings & Landscapes: Journal of the Vernacular Architecture Forum* 16, 2 (2009): 86–11.
Chow, Jeremy, ed. *Eighteenth-Century Humanities*. Rutgers/Bucknell University Press, 2023.
Christie, David. "Nature in Urban Saint John." *Journal of the New Brunswick Museum*. 1978.
Clancy, Peter. *Offshore Petroleum Politics: Regulation and Risk in the Scotian Basin*. UBC Press, 2011.
Clifford, Jim. *West Ham and the River Lea: A Social and Environmental History of London's Industrialized Marshland, 1839–1914*. UBC Press, 2017.
Coates, Peter. *A Story of Six Rivers: History, Culture and Ecology*. Reaktion Books, 2013.
Conrad, Margaret. *At the Ocean's Edge: A History of Nova Scotia to Confederation*. University of Toronto Press, 2020.
Coulin, Nancy, and Carola Hein, eds. *The Urbanisation of the Sea: From Concepts and Analysis to Design*. nai010 Publishers, 2020.

Craig, Béatrice. *Backwoods Consumers and Homespun Capitalists: The Rise of a Market Culture in Eastern Canada.* University of Toronto Press, 2009.

Crutzen, Paul. "Have We Entered the Anthropocene?" *International Geosphere-Biosphere Programme Newsletter* (2000): 17–18.

Curley, Rosemary. "Expansion of Charlottetown Along the Hillsborough Shore." *Island Magazine* 58 (2005): 28–31.

Currie, L., and B. Mair. *Georgetown, Prince Edward Island, 1730–1900.* Georgetown Lions Club, 1982.

Dagenais, Michèle. *Montreal, City of Water: An Environmental History.* Translated by Peter Feldstein. UBC Press, 2017.

Daigle, R. "Sea-Level Rise and Flooding Estimates for New Brunswick Coastal Sections." R.J. Daigle Enviro and the New Brunswick Department of Environment and Local Government, 2020.

Davies, Gwendolyn. "Consolation to Distress: Loyalist Literary Activity in the Maritimes." *Acadiensis* 16, 2 (1987): 51–68.

Davies, Gwendolyn, Peter Larocque, and Christl Verduyn, eds. *The Creative City of Saint John.* Formac Publishing, 2018.

Davies, Jeremy. *The Birth of the Anthropocene.* University of California Press, 2016.

Dawson, Ashley. *Extreme Cities: The Peril and Promise of Urban Life in the Age of Climate Change.* Verso Press, 2017.

De Jong, Nicolas J., and Marven E. Moore. *Shipbuilding on Prince Edward Island: Enterprise in a Maritime Setting, 1787–1920.* Canadian Museum of Civilization, 1994.

Demeritt, David. "Representing the 'True' St. Croix: Knowledge and Power in the Partition of the Northeast." *William & Mary Quarterly* 54, 3 (1997): 515–48.

Desimini, Jill, and Charles Waldheim. *Cartographic Grounds: Projecting the Landscape Imaginary.* Princeton Architectural Press, 2016.

Dryzek, John S., and Jonathan Pickering. *The Politics of the Anthropocene.* Oxford University Press, 2018.

Ducharme, Andre, and Gary Turner, eds. "Preserving the Environment of Halifax Harbour." Workshop 1 and 2. Fisheries and Oceans Canada/Halifax Regional Municipality, 2001.

Dunbar-Hester, Christina. *Oil Beach: How Toxic Infrastructure Threatens Life in the Ports of Los Angeles and Beyond.* University of Chicago Press, 2023.

Edelson, S. Max. *The New Map of Empire: How Britain Imagined America Before Independence.* Harvard University Press, 2017.

Edney, Matthew. *Cartography: The Ideal and Its History.* University of Chicago Press, 2019.

Edney, Matthew. *Mapping an Empire: The Geographical Construction of British India, 1765–1843.* University of Chicago Press, 1997.

Edwards, Justin D., and Douglas Ivison, eds. *Downtown Canada: Writing Canadian Cities.* University of Toronto Press, 2005.

Ermus, Cindy, ed. *Environmental Disaster in the Gulf South: Two Centuries of Catastrophe, Risk, and Resilience.* LSU Press, 2018.

Farell, Barbara, and Aileen Desbarats, eds. *Explorations in the History of Canadian Mapping: A Collection of Essays.* Association of Canadian Map Libraries and Archives, 1988.

Fenech, Adam, and the UPEI Climate Lab. "Keeping Watch over Prince Edward Island's Coastline." *American Review of Canadian Studies* 55, 1 (2025): 87–92.

Fingard, Judith. *The Dark Side of Life in Victorian Halifax*. Pottersfield Press, 1989.

Fingard, Judith, Janet Guildford, and David Sutherland. *Halifax: The First 250 Years*. Formac Publishing, 1999.

Forbes, Ernest R. *The Maritime Rights Movement, 1919–1927: A Study in Canadian Regionalism*. McGill-Queen's University Press, 1979.

Forward, Charles. "The Development of Canada's Five Leading National Ports." *Urban History Review/Revue d'histoire Urbaine* 10, 3 (1982): 25–45.

Frost, James D. *Canada's Atlantic Gateway: An Illustrated History of the Port of Halifax*. Nimbus Publishing, 2008.

Gandy, Matthew. *The Fabric of Space: Water, Modernity, and the Urban Imagination*. MIT Press, 2014.

Gaspar de Freitas, Joana, Robert James, and Isaac Land, "Coastal Studies and Society: The Tipping Point." *Coastal Studies & Society* 1, 1 (2021): 3–9.

Gerald Penney Associates and Newfoundland and Labrador Provincial Archaeology Office. *St. John's Harbour Interceptor Sewer Archaeological Investigations 2004–2010: Technical Report*. Gerald Penney Associates, 2011.

Giblett, Rodney J. *Postmodern Wetlands: Culture, History, Ecology*. Edinburgh University Press, 1996.

Gilliland, Jason, and Mathew Novak. "Positioning the Past with the Present: The Use of Fire Insurance Plans and GIS for Urban Environmental History." *Environmental History* 11, 1 (2006): 136–39.

Gillis, John R. *The Human Shore: Seacoasts in History*. University of Chicago Press, 2013.

Goodell, Jeff. *The Water Will Come: Rising Seas, Sinking Cities, and the Remaking of the Civilized World*. Little, Brown, 2017.

Grant, Jill, and Marcus Paterson. "Scientific Cloak/Romantic Heart: Gordon Stephenson and the Redevelopment Study of Halifax, 1957." *Town Planning Review* 83, 3 (2012): 319–36.

Grant, Jill, and Gladys Wai Kwan Leung. "Disputing the Character of the City: Heritage, Regeneration and the Urban Design Turn." *London Journal of Canadian Studies* 31, 1 (2016): 113–30.

Gray, Steven. *Steam Power and Sea Power: Coal, the Royal Navy, and the British Empire, 1870–1914*. Palgrave Macmillan, 2018.

Greene, Mark A. *The Great Saint John Fire of 1877: The Rise, Destruction and Recovery of Canada's Leading Port City*. Formac Publishing, 2022.

Grydehøj, Adam. "Islands as Legible Geographies: Perceiving the Islandness of Kalaallit Nunaat (Greenland)." *Journal of Marine and Island Cultures* 7, 1 (2018): 1–11.

Guild, Ryan, Xiuquan Wang, and Pedro A. Quijón. "Climate Change Impacts on Coastal Ecosystems." *Environmental Research: Climate* 3, 4 (2024): 1–27.

Gwyn, Julian. *Ashore and Afloat: The British Navy and the Halifax Naval Yard Before 1820*. University of Ottawa Press, 2004.

Hall, Jason. "Maliseet Cultivation and Climatic Resilience on the Wəlastəkw/St. John River During the Little Ice Age." *Acadiensis* 44, 2 (2015): 3–25.

Hall, Jason. "River of Three Peoples: An Environmental and Cultural History of the Wəlastəw/rivière St. Jean/St. John River, c. 1550–1850." PhD diss., University of New Brunswick, 2015.

Haraway, Donna. "Anthropocene, Capitalocene, Plantationocene, Chthulucene: Making Kin." *Environmental Humanities* 6, 1 (2015): 159–65.

Hardenberg, Wilko Graf von. "Knowing the Littoral: Perception and Representation of Terraqueous Spaces in a Global Perspective." *ISIS* 112, 1 (2021): 108–10.

Harley, J.B. *The New Nature of Maps: Essays in the History of Cartography*. Edited by Paul Laxton. Johns Hopkins University Press, 2001.

Harris, R. Cole, ed. *Historical Atlas of Canada*. Vol. 1, *From the Beginning to 1800*. University of Toronto Press, 1987.

Hartog, Hendrik. *Public Property and Private Power: The Corporation of the City of New York in American Law, 1730–1870*. University of North Carolina Press, 1983.

Harvey, David. "Globalization and the 'Spatial Fix.'" *Geographische revue* 3 (2001): 23–30.

Harvey, David. *The Limits to Capital*. Verso, 2006.

Harvey, David, ed. "Limits to Capital." Special issue, *Antipode: A Radical Journal of Geography* 36, 3 (2004).

Harvey, David. "A View from Federal Hill." In *The Baltimore Book: New Views of Local History*. Edited by Elizabeth Fee, Linda Shopes, and Linda Zeidman. Temple University Press, 1991.

Hatvany, Matthew. "'Wedded to the Marshes': Salt Marshes and Economic Differentiation in Early Prince Edward Island." *Acadiensis* 30, 2 (2001): 40–55.

Haumann, Sebastian, Martin Knoll, and Detlev Mares, eds. *Concepts of Urban-Environmental History*. Transcript, 2020.

Hein, Carola, ed. *Adaptive Strategies for Water Heritage: Past, Present and Future*. Springer, 2019.

Hein, Carola. "Planning for Porosity: Exploring Port City Development Through the Lens of Boundaries and Flows." *Urban Planning* 6, 3 (2021).

Hein, Carola, and Dick Schubert. "Introduction: Resilience, Disaster, and Rebuilding in Modern Port Cities." *Journal of Urban History* 47, 2 (2021): 235–49.

Hennessey, Catherine, David L. Keenleyside, and Edward MacDonald, eds. *The Landscapes of Charlottetown*. PEI Museum and Heritage Foundation, 2010.

Hiller, James, and Peter Neary, eds. *Newfoundland in the Nineteenth and Twentieth Centuries: Essays in Interpretation*. University of Toronto Press, 1980.

Holm, Poul, and Verena Winiwarter. "Climate Change Studies and the Human Sciences." *Global and Planetary Change* 156 (2017): 115–22.

Hornby, Jim. *Black Islanders: Prince Edward Island's Historical Black Community*. Institute of Island Studies/University of Prince Edward Island, 1991.

Hornsby, Stephen. *British Atlantic, American Frontier: Spaces of Power in Early Modern British America*. University Press of New England, 2005.

Hornsby, Stephen. *Surveyors of Empire: Samuel Holland, J.F.W. Des Barres, and the Making of the Atlantic Neptune*. McGill-Queen's University Press, 2011.

Hornsby, Stephen, and Richard W. Judd, eds. *Historical Atlas of Maine*. University of Maine Press, 2015.

Hornsby, Stephen, and John Reid, eds. *New England and the Maritime Provinces: Connections and Comparisons*. McGill-Queen's University Press, 2005.

Hubbard, Jennifer, Robert L. Stephenson, and David J. Wildish, eds. *A Century of Maritime Science: The St. Andrews Biological Station*. University of Toronto Press, 2016.

Hunter, R. Alexander. "Cod, Colonialism, and the Anthropocene." *Environment & History* 28, 3 (2022): 369–74.

Hurley, Andrew. *Beyond Preservation: Using Public History to Revitalize Inner Cities*. Temple University Press, 2010.

Inwood, Kris, ed. *Farm, Factory, and Fortune: New Studies in the Economic History of the Maritime Provinces.* Acadiensis Press, 1993.

Jaffee, David. *People of the Wachusett: Greater New England in History and Memory, 1630–1860.* Cornell University Press, 1999.

James, Thomas S., Catherine Robin, Joseph Henton, and M. Craymer. "Relative Sea-Level Projections for Canada Based on the IPCC Fifth Assessment Report and the NAD83v70VG National Crustal Velocity Model." Geological Survey of Canada, Open File 8764, 2021.

Janzen, Olaf. "The Royal Navy and the Defence of Newfoundland During the American Revolution." *Acadiensis* 14, 1 (1984): 28–48. Reprinted in *War and Trade in Eighteenth-Century Newfoundland.* Liverpool University Press, 2013.

Johnson, Alexander. *The First Mapping of America: The General Survey of British North America.* I.B. Tauris, 2017.

Johnston, A.J.B. *Endgame, 1758: The Promise, the Glory, and the Despair of Louisbourg's Last Decade.* University of Nebraska Press, 2007.

Judd, Richard. *Second Nature: An Environmental History of New England.* University of Massachusetts Press, 2014.

Kane, Stephanie. *Where Rivers Meet the Sea: The Political Ecology of Water.* Temple University Press, 2012.

Karrow, P.F., and O.L. White, eds. *Urban Geology of Canadian Cities.* Geological Association of Canada, 1998.

Kiechle, Melanie. *Smell Detectives: An Olfactory History of Nineteenth-Century Urban America.* University of Washington Press, 2017.

Kornwolf, James D., and Georgiana Wallis Kornwolf. *Architecture and Town Planning in Colonial North America.* Vol. 3. Johns Hopkins University Press, 2002.

Land, Isaac. "The Urban Amphibious." In *The New Coastal History: Cultural and Environmental Perspectives from Scotland and Beyond,* edited by David Worthington. Palgrave Macmillan, 2017.

Leeming, Mark. *In Defence of Home Places: Environmental Activism in Nova Scotia.* UBC Press, 2017.

Lees, James. "Uprooting the Forest: Timber Surveying and State Formation in the British Atlantic, 1608–1806." PhD diss., University of Oxford, 2024.

Lennox, Jeffers. "An Empire on Paper: The Founding of Halifax and Conceptions of Imperial Space, 1744–55." *Canadian Historical Review* 88, 3 (2007): 373–412.

Lennox, Jeffers. *Homelands and Empires: Indigenous Spaces, Imperial Fictions, and Competition for Territory in Northeastern North America, 1690–1763.* University of Toronto Press, 2017.

Leroux, John. *Building New Brunswick: An Architectural History.* Goose Lane Editions, 2008.

Levinson, Marc. *The Box: How the Shipping Container Made the World Smaller and the World Economy Bigger.* 2nd ed. Princeton University Press, 2016.

Lewis-Jones, Huw, ed. *The Writer's Map: An Atlas of Imaginary Lands.* University of Chicago Press, 2018.

Li, Yangdong. "Lighting and Temperature in St. John's, Newfoundland." Network in Canadian History and Environment/Nouvelle initiative Canadienne en histoire de l'environnement, 2002. https://niche-canada.org/2022/09/08/lighting-and-temperature-in-the-st-johns/.

Lipman, Andrew. *The Saltwater Frontier: Indians and the Contest for the American Coast.* Yale University Press, 2015.

Liverman, David, Norman Catto, and Martin Batterson. "Geological Hazards and Disasters in St. John's." *Newfoundland & Labrador Studies* 21, 1 (2006): 71–96.

Lockerby, Earle, and Douglas Sobey. *Samuel Holland: His Work and Legacy on Prince Edward Island.* Island Studies Press, 2015.

Loo, Tina. *Moved by the State: Forced Relocation and Making a Good Life in Postwar Canada.* UBC Press, 2019.

Lower, Arthur. *Unconventional Voyages.* Ryerson, 1953.

Luckin, Bill, and Peter Thorsheim, eds. *A Mighty Capital Under Threat: The Environmental History of London, 1800–2000.* University of Pittsburgh Press, 2020.

Lunstrum, Elizabeth, and Pablo S. Bose. "Environmental Displacement in the Anthropocene." *Annals of the American Association of Geographers* 112, 3 (2022): 644–53.

MacDonald, Edward. *If You're Stronghearted: Prince Edward Island in the Twentieth Century.* PEI Museum and Heritage Foundation, 2000.

MacDowell, Lauren. *An Environmental History of Canada.* UBC Press, 2016.

MacEachern, Alan, and George Edward MacDonald. *The Summer Trade: A History of Tourism on Prince Edward Island.* McGill-Queen's University Press, 2022.

MacKinnon, J.B. *The Once and Future World: Nature as It Was, as It Is, as It Could Be.* Houghton Mifflin Harcourt, 2013.

MacKinnon, Wayne, and Elinor Vass. *The Best of the Past: Traditional Sustainable Agriculture in Prince Edward Island.* Institute of Island Studies, 1989.

MacLennan, Hugh. *Barometer Rising.* Duell, Sloan & Pearce, 1941; repr., New Canadian Library edition, McClelland & Stewart, 1989.

MacLennan, Hugh. "An Orange from Portugal." In *An Orange from Portugal: Christmas Stories from the Maritimes and Newfoundland.* Edited by Anne Simpson. Goose Lane Editions, 2003.

MacOwan, B.H. "The Evolution of a Regional Urban Network: New Brunswick and Nova Scotia, 1871–1971." PhD diss., University of Waterloo, 1986.

Marsh, George Perkins. *Man and Nature: Or, Physical Geography as Modified by Human Action.* S. Low, Son and Marston, 1864.

Martin, Ged. "Geography and Governance: The Problem of Saint John (New Brunswick) 1785–1927." https://www.gedmartin.net/martinalia-mainmenu-3/237-geography-and-governance-the-problem-of-saint-john-new-brunswick-1785-1927.

McCalla, Robert J. "Separation and Specialization of Land Uses in Cityport Waterfronts: The Cases of Saint John and Halifax." *Canadian Geographer* 27, 1 (1983): 48–61.

McCann, L.D. "Staples and the New Industrialism in the Growth of Post-Confederation Halifax." *Acadiensis* 8, 2 (1979): 47–79.

McCloskey, J.P. "Bridging the Hillsborough." *Island Magazine* 56 (2004): 32–37.

McGahan, Elizabeth W. *The Port of Saint John.* Vol. 1, *From Confederation to Nationalization, 1867–1927: A Study in the Process of Integration.* National Harbours Board, 1982.

McIlwraith, Thomas. "Digging Out and Filling In: Making Land on the Toronto Waterfront in the 1850s." *Urban History Review/Revue D'histoire Urbaine* 20, 1 (1991): 15–33.

McKay, Ian. *The Quest of the Folk: Antimodernism and Cultural Selection in Twentieth Century Nova Scotia.* McGill-Queen's University Press, 1994.

McKay, Ian, and Robin Bates. *In the Province of History: The Making of the Public Past in Twentieth-Century Nova Scotia.* McGill-Queen's University Press, 2010.

McKenzie, Ruth, ed. *The St. Lawrence Survey Journals of Captain Henry Wolsey Bayfield, 1829–1853*. Publications of the Champlain Society, 1984.

McNeill, J.R. "The Anthropocene and the Eighteenth Century." *Eighteenth-Century Studies* 49, 2 (2016): 117–28.

McNeill, J.R., and Peter Engelke. *The Great Acceleration: An Environmental History of the Anthropocene Since 1945*. Harvard University Press, 2016.

Melosi, Martin. *The Sanitary City: Environmental Services in Urban America from Colonial Times to the Present.* University of Pittsburgh Press, 2008.

Mentz, Steve. *Break Up the Anthropocene*. University of Minnesota Press, 2019.

Moga, Steven. *Urban Lowlands: A History of Neighborhoods, Poverty, and Planning.* University of Chicago Press, 2020.

Montgomery, Lucy Maud. *Anne of Green Gables.* L.C. Page, 1908.

Montgomery, Lucy Maud. *Anne of the Island.* A.L. Burt, 1915.

Montgomery, Lucy Maud. *At the Altar: Matrimonial Tales.* Edited by Rea Wilmshurst. McClelland & Stewart, 1995.

Montgomery, Lucy Maud. *The Complete Journals of L.M. Montgomery, the PEI Years, 1889–1900*. Edited by Mary Rubio and Elizabeth Waterston. Oxford University Press, 2017.

Montgomery, Lucy Maud. *A Name for Herself: Selected Writings, 1891–1917*. Edited by Benjamin Lefebvre. University of Toronto Press, 2018.

Mosley, Stephen. "Coastal Cities and Environmental Change." *Environment & History* 20, 4 (2014): 517–33.

Muise, Del A. "The Federal Election of 1867 in Nova Scotia: An Economic Interpretation." *Collections of the Royal Nova Scotia Historical Society* 36 (1968): 327–51.

Muise, Del A. "'The Great Transformation': Changing the Urban Face of Nova Scotia, 1871–1921." *Nova Scotia Historical Review* 11, 2 (1991): 1–42.

Murray, Jean M., ed. *The Newfoundland Journal of Aaron Thomas, Able Seaman in H.M.S. Boston.* Longman's Green, 1968.

Nader, G.A. *Cities of Canada.* Vol. 2. Macmillan of Canada, 1975–76.

Norman, Howard. *The Museum Guard.* Picador, 1998.

Nova Scotia Department of Environment and Climate Change. *Halifax Census Division: Climate Risk Summary*. December 2022.

Nye, David E. *American Technological Sublime.* MIT Press, 1994.

Olsen, Sherry. "Downwind, Downstream, Downtown: The Environmental Legacy in Baltimore and Montreal." *Environmental History* 12, 4 (2007): 845–66.

O'Neill, Paul. *The Oldest City: The Story of St. John's, Newfoundland.* Press Porcepic, 1975.

Orsi, Jared. *Hazardous Metropolis: Flooding and Urban Ecology in Los Angeles.* University of California Press, 2004.

Parsons, Tom, Pei-Chin Wu, Meng (Matt) Wei, and Steven D'Hondt. "The Weight of New York City: Possible Contributions to Subsidence from Anthropogenic Sources." *Earth's Future* 11 (2023): 1–13.

Pastore, Christopher. *Between Land and Sea: The Atlantic Coast and the Transformation of New England.* Harvard University Press, 2014.

Patterson, Stephen. "Eighteenth-Century Treaties: The Mi'kmaq, Maliseet, and Passamaquoddy Experience." *Native Studies Review* 18, 1 (2009): 25–52.

Pawling, Michah A., ed. *The 1820 Journal and Plans of Survey of Joseph Treat.* University of Massachusetts Press, 2007.

Pawling, Michah A. "Wəlastəkwey (Maliseet) Homeland: Waterscapes and Continuity Within the Lower St. John River Valley, 1784–1900." *Acadiensis* 46, 2 (2017): 5–34.

Payne, Brian. "The Environmental Historiography of the Maritime Peninsula." *Acadiensis* 45, 1 (2016): 163–77.

Peace, Thomas. *The Slow Rush of Colonization: Spaces of Power in the Maritime Peninsula, 1680–1790*. UBC Press, 2023.

Penna, Anthony, and Conrad Wright, eds. *Remaking Boston: An Environmental History of the City and Its Surroundings*. University of Pittsburgh Press, 2009.

Polack, Fiona, and Danine Farquharson, eds. *Cold Water Oil: Offshore Petroleum Cultures*. Routledge, 2022.

Pońka, Anabelle, ed. *Le Port d'Halifax/Halifax Harbour*. Goose Lane/National Gallery of Canada, 2018.

Pope, Peter. *Fish into Wine: The Newfoundland Plantation in the Seventeenth Century*. Omohundro Institute of Early American History and Culture/University of North Carolina Press, 2004.

Pope, Peter. *St. John's Harbour Area Archaeological Potential*. Past Present, Historic Sites and Material Culture Consulting, 1991.

Pope, Peter. *Summer 2000 Excavations at 327 Water St., St. John's (CjAe-08)*. 2003.

Pope, Peter. *Under Water Street: The Early Waterfront of St. John's, Newfoundland*. ICHAM Publications, 1996.

Raddall, Thomas. *At the Tide's Turn and Other Stories*. McClelland & Stewart, 1971.

Raddall, Thomas. *Halifax: Warden of the North*. Doubleday, 1965.

Raddall, Thomas. *In My Time: A Memoir*. McClelland & Stewart, 1976.

Raddall, Thomas. *The Nymph and the Lamp*. Little, Brown, 1950.

Rawson, Michael. *Eden on the Charles: The Making of Boston*. Harvard University Press, 2010.

Rees, Ronald. *St. Andrews and the Islands*. Nimbus Publishing, 1995.

Reid, John. "Empire, the Maritime Colonies, and the Supplanting of Mi'kma'ki/Wulstukwik, 1780–1820." *Acadiensis* 38, 2 (2009): 78–97.

Remes, Jacob A.C. *Disaster Citizenship: Survivors, Solidarity, and Power in the Progressive Era*. University of Illinois Press, 2016.

Reuss, Martin, and Stephen H. Cutcliffe, eds. *The Illusory Boundary: Environment and Technology in History*. University of Virginia Press, 2010.

Reynolds, Tyler. "Re-imagining a Petrochemical Landscape in Dartmouth Nova Scotia." MArch thesis, Dalhousie University, 2016.

Rider, Peter. *Charlottetown: A History*. Canadian Museum of Civilization/PEI Museum and Heritage Foundation, 2009.

Roney, Jessica Choppin, ed. "Port Cities in the Early Modern World, 1500–1800." Special issue, *Early American Studies* 15, 4 (2017).

Rubin, M. Jasper. *A Negotiated Landscape: The Transformation of San Francisco's Waterfront Since 1950*. University of Pittsburgh Press, 2016.

Rudin, Ronald. *Against the Tides: Reshaping Landscape and Community in Canada's Maritime Marshlands*. UBC Press, 2021.

Ruffman, Alan, and Colin D. Howell, eds. *Ground Zero: A Reassessment of the 1917 Explosion in Halifax Harbour*. Nimbus Publishing/Gorsebrook Research Institute for Atlantic Canada Studies, 1994.

Rutland, Ted. *Displacing Blackness: Power, Planning, and Race in Twentieth-Century Halifax*. University of Toronto Press, 2018.

Ryan, Mary. *Taking the Land to Make the City: A Bicoastal History of North America.* University of Texas Press, 2019.
Ryan, Shannon. "Fishery to Colony: A Newfoundland Watershed, 1793–1815." *Acadiensis* 12, 2 (1983): 34–52.
Sachs, Aaron. *Arcadian America: The Death and Life of an Environmental Tradition.* Yale University Press, 2013.
Sackett, Andrew. "Inhaling the Salubrious Air: Health and Development in St. Andrews, N.B., 1880–1910." *Acadiensis* 25, 1 (1995): 54–81.
Sager, Eric, and Gerald E. Panting. *Maritime Capital: The Shipping Industry in Atlantic Canada, 1820–1914.* McGill-Queen's University Press, 1990.
Samson, Daniel. *The Spirit of Industry and Improvement: Liberal Government and Rural-Industrial Society, Nova Scotia, 1790–1867.* McGill-Queen's University Press, 2008.
Sandwell, R.W., ed. *Powering Up Canada: The History of Power, Fuel, and Energy from 1600.* McGill-Queen's University Press, 2016.
Sarty, Roger, and Doug Knight. *Saint John Fortifications, 1630–1956.* Goose Lane Editions, 2003.
Scanlon, Joseph, and Roger Sarty. *Catastrophe: Stories and Lessons from the Halifax Explosion.* Wilfrid Laurier University Press, 2020.
Scarth, Kate. "Born for City Life? L.M. Montgomery's Urban Prince Edward Island." *Island Magazine* 86 (2020): 50–60.
Schlichting, Kara. *New York Recentered: Building the Metropolis from the Shore.* University of Chicago Press, 2019.
Schwinghamer, Steven. "'Altogether Unsatisfactory': Revisiting the Opening of the Immigration Facility at Halifax's Pier 21." *Journal of the Royal Nova Scotia Historical Society* 15 (2012): 61–74.
Schwinghamer, Steven, and Jan Raska. *Pier 21: A History.* University of Ottawa Press, 2020.
Seager, Allen. "Minto, New Brunswick: A Study in Canadian Class Relations Between the Wars." *Labour/Le Travail* 5 (1980): 81–132.
Seasholes, Nancy. *Gaining Ground: A History of Landmaking in Boston.* MIT Press, 2003.
Settle, Victor L. "Halifax Shipyards, 1918–1978: An Historical Perspective." MA thesis, Saint Mary's University, 1994.
Sharpe, Errol. *A People's History of Prince Edward Island.* Steel Rail Publishing, 1976.
Shears, Robert. "Examination of a Contested Landscape: Archaeological Prospection on the Eastern Shore of Nova Scotia." MA Thesis, St. Mary's University, 2013.
Sieber, R. Timothy. "Waterfront Revitalization in Postindustrial Port Cities of North America." *City & Society* 5, 2 (1991): 120–36.
Šimková, Pavla. *Urban Archipelago: An Environmental History of the Boston Harbor Islands.* University of Massachusetts Press, 2021.
Skelton, Leona. *Tyne After Tyne: An Environmental History of a River's Battle for Protection 1529–2015.* White Horse Press, 2017.
Slaymaker, Olav, and Norman Catto, eds. *Landscapes and Landforms of Eastern Canada.* Springer, 2020.
Smith, Carl. *City Water, City Life: Water and the Infrastructure of Ideas in Urbanizing Philadelphia, Boston, and Chicago.* University of Chicago Press, 2014.
Smith, Harry, and Maria Soledad Garcia Ferrari, eds. *Waterfront Regeneration: Experiences in City-Building.* Routledge, 2012.

Smith, Joshua. *Borderland Smuggling: Patriots, Loyalists, and Illicit Trade in the Northeast, 1783–1820.* University Press of Florida, 2006.

Smith, Marilyn G. *The King's Yard: An Illustrated History of the Halifax Dockyard.* Nimbus Publishing, 1985.

Smitheram, Verner, David Milne, and Satadal Dasgupta, eds. *The Garden Transformed: Prince Edward Island, 1945–1980.* Ragweed Press, 1982.

St. Andrews, N.B.: Miscellaneous Reports. Manuscript Report Number 134. Parks Canada, 1965–66.

Stein, Blair. "Environmental History at Work: New Environmental Histories of Canada and Atlantic Canada." *Acadiensis* 50, 1 (2022): 135–48.

Steinberg, Phillip E. "Of Other Seas: Metaphors and Materialities in Maritime Regions." *Atlantic Studies* 10, 2 (2013): 156–69.

Steinberg, Phillip E. *The Social Construction of the Ocean.* Cambridge University Press, 2001.

Steinberg, Ted. *Gotham Unbound: The Ecological History of Greater New York.* Simon & Schuster, 2015.

Stoll, Mark. *Profit: An Environmental History.* Polity Press, 2023.

Sullivan, Cheryl. "The Paradox of Bluenose II: Antimodernism, Capitalism and the Legacy of the Schooner Bluenose in Nova Scotia." *Nova Scotia Historical Review* 16, 2 (1996): 1–22.

Sullivan, David, ed. *St. Andrews: An Historical Scrapbook.* Pendlebury Press, 2008.

Sutherland, David. "Halifax 1815–1914: 'Colony to Colony.'" *Urban History Review/Revue d'histoire Urbaine* 4, 1 (1975): 7–11.

Sutherland, David. *"We Harbor No Evil Design": Rehabilitation Efforts After the Halifax Explosion of 1917.* The Champlain Society, University of Toronto Press, 2017.

Sutter, Paul, and Paul M. Pressley, eds. *Coastal Nature, Coastal Culture: Environmental Histories of the Georgia Coast.* University of Georgia Press, 2018.

Szeman, Imre, and Jeff Diamanti, eds. *Energy Culture: Art and Theory on Oil and Beyond.* West Virginia University Press, 2019.

Tarr, Joel, ed. *Devastation and Renewal: An Environmental History of Pittsburgh and Its Region.* University of Pittsburgh Press, 2003.

Taylor, Graham D. *Imperial Standard: Imperial Oil, Exxon, and the Canadian Oil Industry from 1880.* University of Calgary Press, 2019.

Taylor, M. Brook. *A Camera on the Banks: Frederick William Wallace and the Fishermen of the Grand Banks.* Goose Lane Editions, 2000.

Thomas, Julia Adeney. *The Historian's Task in the Anthropocene.* Princeton University Press, forthcoming.

Tuck, Robert C. "Georgetown: The Town That Time Forgot." *The Island Magazine* 4 (1978): 22–28.

Tulloch, Tom. "The Halifax Graving Dock and the 1917 Explosion." *Argonauta* 36, 1 (2019): 6–22.

Tyszczuk, Renata. *Provisional Cities: Cautionary Tales for the Anthropocene.* Routledge, 2019.

Van Duzer, Chet, and Lauren Beck. *Canada Before Confederation: Maps at the Exhibition.* Vernon Press, 2017.

Vileisis, Ann. *Discovering the Unknown Landscape: A History of America's Wetlands.* Island Press, 1997.

Wade, Tracey, and CLIMAtlantic. "Sea Level Rise and Public Health Implications." National Collaborating Centre for Environmental Health, 2022.

Wallace, C.M. "Saint John, New Brunswick (1800–1900)." *Urban History Review* 4 (1975): 12–21.

Wallace, C.M. "Saint John Boosters and the Railroads in Mid-nineteenth Century." *Acadiensis* 6, 1 (1976): 71–91.

Wallace, Frederick William. *The Romance of a Great Port.* Barnes, 1935.

Watson, Andrew. *Making Muskoka: Tourism, Rural Identity, and Sustainability, 1870–1920.* UBC Press, 2023.

Way, Thaïsa, ed. *River Cities: City Rivers.* Dumbarton Oaks Research Library and Collection, 2018.

"What Is the Anthropocene? – Current Definition and Status." Subcommission on Quaternary Stratigraphy, Anthropocene Working Group. Accessed 9 Janaury 2026. https://quaternary.stratigraphy.org/working-groups/anthropocene.

Wilkinson, Bruce. "Humans as Geologic Agents: A Deep-Time Perspective." *Geology* 33, 3 (2005): 161–64.

Williams, Ron. *Landscape Architecture in Canada.* McGill-Queen's University Press, 2014.

Wilson, Sheena, Adam Carlson, and Imre Szeman, eds. *Petrocultures: Oil, Politics, Culture.* McGill-Queen's University Press, 2017.

Winiwarter, Verena, Gertrud Haidvogl, Severin Hohensinner, Friedrich Hauer, and Michael Bürkner. "The Long-Term Evolution of Urban Waters and Their Nineteenth Century Transformation in European Cities. A Comparative Environmental History." *Water History* 8, 3 (2016): 209–33.

Worster, Donald. "Ice, Worms, and Dirt: The Power of Nature in North American History." In *Method and Meaning in Canadian Environmental History,* edited by Alan MacEachern and William J. Turkel. Nelson, 2009.

Wynn, Graeme. "Thinking About Mountains, Valleys and Solitudes: Historical Geography and the New Atlantic History." *Acadiensis* 31, 1 (2001): 129–45.

Wynn, Graeme. *Timber Colony: A Historical Geography of Early Nineteenth-Century New Brunswick.* University of Toronto Press, 1981.

Wynn, Graeme. "'Tracing One Warm Line Through a Land so Wide and Savage.' Fifty Years of Historical Geography in Canada." *Historical Geography* 40 (2012): 5–32.

Wynn, Graeme. "W.F. Ganong, A.H. Clark and the Historical Geography of Maritime Canada." *Acadiensis* 10, 2 (1981): 5–28.

Young, Alan R., ed. *Time and Place: The Life and Works of Thomas H. Raddall.* Acadiensis Press, 1991.

Index

Note: Page numbers with (f) refer to illustrations.

NATURE | HISTORY | SOCIETY

Claire Elizabeth Campbell, *Shaped by the West Wind: Nature and History in Georgian Bay*
Tina Loo, *States of Nature: Conserving Canada's Wildlife in the Twentieth Century*
Jamie Benidickson, *The Culture of Flushing: A Social and Legal History of Sewage*
John Sandlos, *Hunters at the Margin: Native People and Wildlife Conservation in the Northwest Territories*
William J. Turkel, *The Archive of Place: Unearthing the Pasts of the Chilcotin Plateau*
Greg Gillespie, *Hunting for Empire: Narratives of Sport in Rupert's Land, 1840–70*
James Murton, *Creating a Modern Countryside: Liberalism and Land Resettlement in British Columbia*
Stephen J. Pyne, *Awful Splendour: A Fire History of Canada*
Hans M. Carlson, *Home Is the Hunter: The James Bay Cree and Their Land*
Sharon Wall, *The Nurture of Nature: Childhood, Antimodernism, and Ontario Summer Camps, 1920–55*
Joy Parr, *Sensing Changes: Technologies, Environments, and the Everyday, 1953–2003*
Liza Piper, *The Industrial Transformation of Subarctic Canada*
Jamie Linton, *What Is Water? The History of a Modern Abstraction*
Dean Bavington, *Managed Annihilation: An Unnatural History of the Newfoundland Cod Collapse*
J. Keri Cronin, *Manufacturing National Park Nature: Photography, Ecology, and the Wilderness Industry of Jasper*
Shannon Stunden Bower, *Wet Prairie: People, Land, and Water in Agricultural Manitoba*
Jocelyn Thorpe, *Temagami's Tangled Wild: Race, Gender, and the Making of Canadian Nature*
Darcy Ingram, *Wildlife, Conservation, and Conflict in Quebec, 1840–1914*
Sean Kheraj, *Inventing Stanley Park: An Environmental History*
Caroline Desbiens, *Power from the North: Territory, Identity, and the Culture of Hydroelectricity in Quebec*
Daniel Macfarlane, *Negotiating a River: Canada, the US, and the Creation of the St. Lawrence Seaway*
Justin Page, *Tracking the Great Bear: How Environmentalists Recreated British Columbia's Coastal Rainforest*

Ryan O'Connor, *The First Green Wave: Pollution Probe and the Origins of Environmental Activism in Ontario*

John Thistle, *Resettling the Range: Animals, Ecologies, and Human Communities in British Columbia*

Jessica van Horssen, *A Town Called Asbestos: Environmental Contamination, Health, and Resilience in a Resource Community*

Nancy B. Bouchier and Ken Cruikshank, *The People and the Bay: A Social and Environmental History of Hamilton Harbour*

Carly A. Dokis, *Where the Rivers Meet: Pipelines, Participatory Resource Management, and Aboriginal-State Relations in the Northwest Territories*

Jonathan Peyton, *Unbuilt Environments: Tracing Postwar Development in Northwest British Columbia*

Mark R. Leeming, *In Defence of Home Places: Environmental Activism in Nova Scotia*

Jim Clifford, *West Ham and the River Lea: A Social and Environmental History of London's Industrialized Marshland, 1839–1914*

Michèle Dagenais, *Montreal, City of Water: An Environmental History*

David Calverley, *Who Controls the Hunt? First Nations, Treaty Rights, and Wildlife Conservation in Ontario, 1783–1939*

Jamie Benidickson, *Levelling the Lake: Transboundary Resource Management in the Lake of the Woods Watershed*

Daniel Macfarlane, *Fixing Niagara Falls: Environment, Energy, and Engineers at the World's Most Famous Waterfall*

Angela V. Carter, *Fossilized: Environmental Policy in Canada's Petro-Provinces*

Stéphane Castonguay, *The Government of Natural Resources: Science, Territory, and State Power in Quebec, 1867–1939*

Ronald Rudin, *Against the Tides: Reshaping Landscape and Community in Canada's Maritime Marshlands*

Andrew Watson, *Making Muskoka: Tourism, Rural Identity, and Sustainability, 1870–1920*

Shannon Stunden Bower, *Transforming the Prairies: Agricultural Rehabilitation and Modern Canada*

Set in Garamond and Charis by Artegraphica Design Co. Ltd.

Copyeditor: Candida Hadley

Proofreader: Judith Earnshaw

Indexer: Judy Dunlop

Cover designer: George Kirkpatrick

Cover image: J.S. Clow, *View of Halifax in Nova Scotia* (1840), Library and Archives Canada, 1970-188-1985

Authorized Representative: Easy Access System Europe – Mustamäe tee 50, 10621 Tallinn, Estonia, gpsr.requests@easproject.com

FSC
www.fsc.org
MIX
Papier | Fördert gute Waldnutzung
FSC® C083411

Kolibri 360 GmbH
Ferdinand-Jühlke-Straße 7
99095 Erfurt, Deutschland
produktsicherheit@kolibri360.de